RAPID PROPAGATION OF FAST-GROWING WOODY SPECIES

CASAFA Report Series

Report Series of CASAFA (Committee on the Application of
Science to Agriculture, Forestry and Aquaculture) of the
International Council of Scientific Unions (ICSU), published
by CAB International:

Rapid Propagation of Fast-growing Woody Species

Edited by

F.W.G. Baker

*Scientific Secretary
CASAFA*

C·A·B International

for

CASAFA

(*Committee on the Application of Science to Agriculture, Forestry and Aquaculture*)

C·A·B International   Tel: Wallingford (0491) 32111
Wallingford   Telex: 847964 (COMAGG G)
Oxon OX10 8DE   Telecom Gold/Dialcom: 84: CAU001
UK   Fax: (0491) 33508

A catalogue record for this book is available from the British Library

ISBN 0 85198 742 7

Typeset by Leaper & Gard Ltd, Bristol
Printed and bound in the UK by Redwood Press Ltd, Melksham

Contents

Preface

A number of studies and programmes have focused attention on tropical and to a lesser extent temperate forest ecosystems and their mismanagement and overexploitation. These include the report of the World Commission on Environment and Development; the Tropical Forestry Action Plan/Programme co-ordinated by the UN Food and Agriculture Organization; the Netherlands Tropenbos Programme, and so on. These give various and varying figures for the rates of removal of forest cover.

Recent figures from FAO indicate that the rate of tropical deforestation has increased from 11.3 million hectares a year in 1980 to 17 million hectares a year in 1990. In addition to the efforts to conserve tropical and other forests efforts are being made to replant as much new forest as possible. The manual and mechanical problems of replanting tree seedlings over areas as large as 10 million hectares are compounded by the difficulties of obtaining or producing the millions of seedlings required for such a planting programme.

These difficulties stimulated CASAFA to organize a symposium on the rapid and mass propagation of woody species at the Secretariat of the International Council of Scientific Unions (ICSU) in late 1989. The symposium brought together a group of scientists with experience in the rapid and mass propagation of a number of tropical and temperate species including *Acacia, Casuarina, Eucalyptus, Musa, Paulownia, Populus, Salix* and bamboos. This third volume in the CASAFA Report Series gives the papers presented and the Recommendations.

The purpose of the ICSU Committee on the Application of Science to Agriculture, Forestry and Aquaculture is to bring together in common purpose scientists engaged in applied research with those with complementary fundamental research in the biological and physical sciences.

CASAFA's principal objective is to advance progress in agriculture, forestry and aquaculture by the more rapid application of basic science to applied problems. The Committee also seeks to improve co-operation between scientists in developed and less developed countries.

The members of CASAFA are grateful to the Australian Commonwealth Scientific and Industrial Research Organization (CSIRO), the Canadian International Development Research Centre (IDRC), the International Network for the Improvement of Banana and Plantain (INIBAP), the German Federal Ministry for Economic Co-operation (BMZ) and the French Centre Technique Forestier Tropical (CTFT) for their support and contributions to the Symposium and for a visit to its research facilities at Nogent sur Marne.

F.W.G. Baker
Scientific Secretary, CASAFA

Chapter 1

Plant Cell and Tissue Culture: Progress and Prospects

Joseph H. Hulse

Chairman, CASAFA

The term 'plant tissue culture' designates almost all forms of plant cultures grown *in vitro*, ranging over single cells, unorganized groups of cells (callus), to highly organized multi-cellular masses (tissue) and organ cultures. In ideal systems, by control of the biochemical and physical environment, cultured plant cells are induced first to multiply then to regenerate into whole plants that can survive transplantation into the greenhouse or field. Cell and tissue culture offers the potential to reduce a plant species or any part of it to a single cell or group of cells; of which as few as five or as many as 500 million can be cultured in a single flask. The ability to select and regenerate single totipotent cells into whole plants offers notable opportunities for the rapid and large scale propagation of many species. The expanding capability to generate some species from single protoplasts, cells from which the walls have been enzymatically removed, and from the fused protoplasts of two different species provide plant breeders with novel sources of genetic diversification and propagation.

Tissue culture has been known since at least the start of this century when an Austrian, Haberlandt, recorded his attempts to regenerate whole plants from single cells. The successful development of artificial growth media for *Nicotiana* species, the long-term culture of carrot tissue in the 1930s, and the later discovery of the influence of auxins, cytokinins and other growth regulators inspired confidence that *in vitro* culture could blossom into more than an academic curiosity.

The successes of plant breeders at IRRI (Philippines) and CIMMYT (Mexico) in generating rice and wheat genotypes with high yield potentials stimulated a significant increased investment in crops research, particularly through the Consultative Group on International Agricultural Research

(CGIAR) consortium of donors which finances the family of International Agricultural Research Centres (IARC). The CGIAR began in 1971 with four centres and a budget of US$12m. Its present annual budget is of the order of US$250m which supports some 15 centres and related activities, and which is exploring how best it can expand and diversify in the future.

Crop improvement by conventional breeding is essentially slow and some of the major cereals seem to be reaching yield plateaux. Higher yield potential is not necessarily concomitant with greater resistance to pathogens, pests and parasites, many of which are adept at evolving resistance or immunity to chemical pesticides. Biological control, which employs natural or exotic enemies to attack unwanted pests, calls for skill and experience in morphological and biochemical taxonomy to classify accurately each pest, and in entomological, mycological and parasitological methodologies to identify, propagate and safely release effective enemies. The environmental hazards that go with high levels of chemical pesticides and fertilizers are a further deterrent to this means of obtaining high levels of plant production and protection.

Consequently the prospect of breeding more efficient plants with built-in resistance to pests and pathogens is appealing to plant scientists, environmentalists, government and donor agencies alike. Wide interspecific hybridization using cell and tissue techniques appear as possible means to these ends. Sixty-nine countries are members of the International Association of Plant Tissue Culture, and the World Bank in co-operation with ISNAR (International Service for National Agricultural Research) and ACIAR (Australian Centre for International Agricultural Research) has recently published a summary account of some of their activities. A review of their reports shows that all the IARCs responsible for crop improvement are engaged in cell and tissue culture both in-house and through collaboration with external co-operating partners.

Plant cell and tissue culture for woody perennials

Over many years plant breeding has progressively benefited from cell biology research. Genetic manipulations by plant cell and tissue culture (PCTC) can provide such new breeding materials and genetic and cytogenetic variants; haploids; disease free plants propagated out of explants from infected cultivars; wide-cross hybrids, together with large numbers of propagules from limited parental stock. PCTC offers a promising bridge between molecular genetic transformations and plant breeding since at the present state of the art, gene, chromosome and organelle transfers need the intermediary of a cultured cell or tissue of the recipient species.

As stated above, PCTC can be achieved through the following.

1. Callus culture, in which masses of unorganized cell clusters are grown on nutrient agar; the explants from which callus derives include embyronic, seedling and mature vegetative parts of such reproductive tissue as ovules.

2. Cell suspension culture in liquid nutrient media, agitated to aerate and break up cell clusters; starting materials may include callus or wounded vegetable tissue.

3. Organ culture from excisions of roots, shoots, embryos, anthers, ovaries or ovules inoculated on to nutrient agar.

4. Meristem tip culture in which meristemic tissue inoculated on to nutrient agar generates single plants or numbers of shoot buds.

5. Protoplast culture: the protoplasts being isolated from leaf mesophyll or root tissue, or from cell suspension cultures by enzymatic digestion followed by culture in liquid media to achieve cell wall regeneration and cell division.

In most culture processes, one or more transfers through several different media may be needed before intact plants with viable balanced root and shoot systems are realized. The media composition will vary according to species and type of culture. Media ingredients usually include: a carbon energy source, such as sucrose, inorganic salts, phytohormones, growth regulators, other nutrients such as vitamins, coconut milk and casein.

A major constraint to orderly progress in *in vitro* culture is inadequate reliable quantitative physiological, biochemical and cell biological data relative to the species to be cultured; together with an understanding of the critical functions of phytohormones, plant growth regulators and other critical media constituents. Consequently PCTC relies more upon empirical experience than fundamental principles. Culture systems that work for one species or explant are often not readily adaptable to others. While callus, suspension and meristem cultures have been demonstrated with many hundreds of plants, regeneration of whole intact plants often proves more difficult. To be of practical and economic benefit, transformed and selected variant cells, cultured products of heterokaryon fusion and haploid micro-spores must regenerate into whole plants which, in turn, generate seed or vegetative progeny that embody and express the genetic traits desired. Experience suggests it may take several years for traits transferred through PCTC to stabilize in an acceptable genetic background and to be consist-ently inherited by subsequent progeny. Significant variability in response may be expected among different species. Nevertheless, PCTC offers unique techniques to plant breeders, techniques that exceed the potential of traditional practices.

Applications of cell and tissue culture

PCTC is applicable to various objectives, several of which overlap or are complementary. These include:

1. clonal propagation;
2. provision of haploids and homozygous lines;
3. control of disease;
4. storage and movement of germplasm;
5. transgenesis and wide interspecific hybridization;
6. generation and selection of genetic variants.

Clonal propagation

Clonal propagation (CP) permits production of large numbers of unique genotypes and is particularly useful where seed production is difficult; to maintain specific rare sterile or sexually incompatible species. The numbers of plant species from which buds, shoots, embryoids and intact plants have been cultured *in vitro* indicates the extent of interest and potential for clonal propagation.

Though CP may appear uneconomic for agronomic crops that can be readily propagated from seed, considerable optimism is expressed for the future of CP applied to woody perennials. Species of the following genera have reportedly been induced to produce multiple shoots and roots *in vitro*: *Malus, Prunus, Vitis, Rubus, Ribes, Actinidia, Morus* and *Vaccinium.* The advantages of CP over nursery propagation for woody species include: independence of season and climate, and improved phytosanitary control since CP takes place within a controlled environment; particular suitability to species that require long generation times; and the ability to produce large numbers of selected or elite propagules for extensive reforestation programmes.

Disease control

PCTC can be adapted to provide pathogen-free plants and thus permit safe phytosanitary control in international germplasm transport. Elimination of pathogens also improves yield and other characters. Fungal and bacterial infections may be controlled by surface sterilization of explants. Intracellular viruses are generally less frequently found in meristemic cells, hence meristem tip culture is effective in generating pathogen-free propagules from infected plants.

Germplasm storage

Though seed is the most convenient form of germplasm for storage and transport, plants propagated vegetatively may be most economically conserved by tissue culture. CIAT's cassava tissue culture bank can accommodate 6000 accessions in a space 7×6×2.5 m. A field gene bank of equivalent capacity would cover approximately 13 ha. The International Board for Plant Genetic Resources (IBPGR) report satisfactory survival of several *Pomus* and *Prunus* species in tissue culture though clearly there is need for considerably more research. The USDA has compiled a comprehensive inventory of fruit and nut tree genetic resource collections in North America and Europe. It seems unlikely, however, that all the named collections are permanently maintained or systematically classified, using, for example, descriptors such as IBPGR has published and recommended for citrus species.

Wide hybridization

A superficial review of the literature suggests a greater investment of time and effort in pursuit of wide interspecific hybrids among annual food crops than woody perennials. All of the International Agricultural Research Centres are pursuing wide interspecific hybrids mainly to transfer into cultivated species from distantly related wild species the latters' evolved resistances to various infections and infestations. To these ends most progress has resulted from embryo rescue techniques though there is increasing interest in somatic hybridization through protoplast fusion.

Variant selection

Regeneration of genetic variants via organogenesis or somatic embryogenesis is reported by many investigators. Genetic variants result from protoplast and cell suspension cultures, some with, some without stimulation by chemical or radiation mutagens. Variants are described that variously show morphological, physiological, nuclear, chromosomal and cytoplasmic differences from the parent plants. Again, most such reports relate to annual agronomic crops.

PCTC of woody perennials

The following is a very incomplete random sample of some issues of interest culled from recent publications.

1. Several authors report albinism, chlorophyll mutants and absence of photosynthesis in shoot cultures of various woody species.

2. Others report as most promising:

(a) multiple shoot cultures, the harvested shoots being induced to root by tranfer to other media;

(b) culture of explants from terminal and axillary buds, and from flower buds containing vegetative primordia;

(c) explants excised during growth flush rather than from dormant tissue.

3. In several instances high frequency rooting appeared more difficult than shoot multiplication.

4. Some reports of shoot and embryo differentiation from callus derived from ovules and immature seeds.

5. Several fruiting species, including raspberry and loganberry have been propagated by shoot tip culture.

The literature contains many formulae and recommendations for culture media including the following.

1. Cytokinins are beneficial in most shoot cultures; auxins appear to be synthesized in some cultures of shoots and young leaf explants.

2. Phenolics including (−)epicatechin, leucocyanidin and quercitin stimulated root development in embryo cultures.

3. Phloroglucinol increased the number, length and weight of cultured apple shoots.

4. Influential variability among agar samples from different sources.

5. Comparisons among culture media are more reliable when constituent concentrations are expressed in molar units than by weight.

In summary, and taking into account the far larger numbers of woody perennials of economic interest, the literature relating to PCTC deals predominantly with agronomic annuals. The opportunities for tree species remain largely unexplored.

Chapter 2

The Role of Micropropagation for Australian Tree Species

V.J. Hartney and J.G.P. Svensson

CSIRO Division of Forestry, PO Box 4008, Canberra, ACT 2600, Australia

Micropropagation has an important role in the rapid multiplication of elite genotypes (e.g. clones and families) of tree species. This is particularly true in tropical and sub-tropical countries where many species have only recently been domesticated, high growth rates are possible, and the demand for wood is very high.

Micropropagation can readily fit into tree improvement programmes even where the trees planted in the forest are seedlings or cuttings.

Research at the CSIRO Division of Forestry has concentrated on developing commercial micropropagation techniques for the rapid multiplication of *Eucalyptus*, *Melaleuca*, *Acacia* and some other Australian species. Simple techniques have been developed that enable propagation of many important species, using minimum laboratory and nursery facilities. These methods will be discussed in this chapter. Such methods need to be tested in other countries to exploit the unique advantages that micropropagation has over alternative techniques such as seedlings and cuttings.

The unique advantages of micropropagation are firstly, a much higher multiplication rate compared with propagation by cuttings, and secondly, the ability to maintain and exchange disease-free genotypes *in vitro.*

For far too long micropropagation has been regarded as being too expensive and has not been considered a viable option in tree improvement programmes. The importance of micropropagation in the improvement of banana, plantain and bamboos (see other chapters in this volume), and a wide range of horticultural crops illustrate the point that all methods of propagation can be utilized for plant improvement and domestication, even with minimal facilities.

Australia is fortunate in having a large number of tree species adapted to a wide range of environments. There are over 500 species of *Eucalyptus,* over 1000 *Acacia* species, about 60 casuarinas and over 150 *Melaleuca* species (Boland *et al.*, 1984). This illustrates the wide genetic diversity at the species level in some of the most important genera.

The domestication of these species is in its infancy in most countries. In many cases we are only just beginning to exploit genetic variation at the species and provenance level. This fact, combined with the large amount of genetic variation often found means that vegetative propagation (by cuttings or micropropagation) allows breeders to make large gains in each cycle of selection. Figure 2.1 illustrates some of the ways that vegetative propagation can be utilized in a tree improvement programme.

After several years, trees growing in well designed field trials are selected for desirable characteristics such as a high growth rate, disease resistance or high pulp yield. Vegetative propagation of selected genotypes (e.g. selected clones or families) for the establishment of a plantation is a much faster means of implementing tree improvement than producing improved seed through a seed orchard or a controlled pollination programme (Matheson and Lindgren, 1985). In addition, it is not uncommon to achieve gains of at least double that obtained from routine plantations of seedlings if clones of the best selections are grown (Brandão, 1984; Turnbull *et al.*, 1988; Zobel *et al.*, 1987).

However, it must be emphasized that propagation of trees by any means is only one part of a tree improvement programme. The essential steps in tree improvement of producing genetic variation and then selecting among genetic variants is an ongoing and continuous process. Vegetative propagation enables us to incorporate our best genotypes into our production forests more rapidly. Many Australian tree species can be micropropagated (Table 2.1; Hartney, 1982; McComb and Bennett, 1986). However, not all species can be readily propagated by micropropagation or by cuttings. Important pulpwood species like *Eucalyptus nitens* and *E. globulus* are notoriously difficult to propagate reliably by cuttings or by micropropagation. However, for all species that we have studied there is

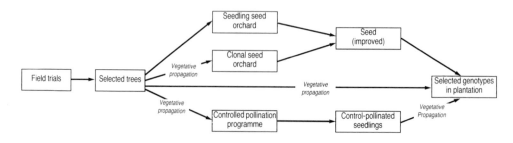

Fig. 2.1. A scheme to illustrate the role of vegetative propagation in tree improvement.

Table 2.1. Number of clones of various species micropropagated under various conditions.

	Under commercial conditions	Under research conditions	Presently recalcitrant	Rooted *ex vitro*[*]
Acacia auriculiformis	4	3	1	
A. maconochieana	1			
A. mangium	6	5		
A. stenophylla	2			
Allocasuarina verticilliata	1			y
Casuarina glauca			1	
Chrysanthemum cinerariifolium (Pyrethrum)	2			y
Eucalyptus aggregata	1			
E. andrewsii subsp. *campanulata*			1	
E. annulata	1			
E. blakelyi		2	1	
E. camaldulensis	42	11		y
E. curtisii	1			
E. deglupta	1			y
E. desmondensis	1			
E. diptera	1		1	
E. ficifolia		1		
E. grandis	4	3		
E. gillii	1			
E. globulus subsp. *bicostata*		1	1	
E. globulus subsp. *globulus*	1	6	2	y
E. macrandra	1			
E. marginata	6			
E. melliodora		1		
E. nitens		20	2	
E. nites × *globulus*	2	1		
E. ochrophloia		1	1	
E. ovata		1	1	
E. pileata	1			
E. rudis	1			
E. tereticornis	1			y
E. viridis	1			
E. wandoo		4	5	
E. yarraensis		1		y
Flindersia brayleana		2		
Melaleuca alternifolia	7			y
M. bracteata	4	1		y
M. cajuputi		1		
M. decora			1	

Table 2.1. Continued

	Under commercial conditions	Under research conditions	Presently recalcitrant	Rooted *ex vitro**
Melaleuca eleuterostachya			1	
M. glomerata	1	2		
M. halmaturorum	2	2	1	
M. lanceolata	1	3	1	
M. lateriflora subsp. *lateriflora*	4	1		
M. quinquenervia			1	
M. thyoides	3	2		
Pinus radiata		20		y
Populus deltoides		2		y
P. deltoides × *P. nigra*		3		y
Simmondsia chinensis (Jojoba)		2		
Toona australis		1		
Totals	105	103	22	

*y = successfully rooted *ex vitro*

always a strong correlation between their ability to be propagated by cuttings and by micropropagation. Species that can be readily propagated by cuttings (e.g. *E. camaldulensis*, *E. grandis*, *E. tereticornis* and various eucalypt hybrids) can also be readily micropropagated.

Many of the species listed in Table 2.1 were selected for their tolerance to salinity by growing them under greenhouse conditions and irrigating the seedlings with a nutrient solution in which the concentration of salt was gradually increased (Aswathappa *et al.*, 1987; Marcar, 1989). Seedlings which survived salinity levels above that of sea water were vegetatively propagated by micropropagation (Hartney and Kabay, 1984).

Extreme caution must be exercised in extrapolating from performance under greenhouse conditions to performance in the field. Salt tolerant trees may grow slowly, be eaten by insects, suffer from disease or die in the next drought. Field performance, after several years, must be the ultimate criterion for evaluation of genotypes. Some of the claims made about these salt tolerant clones are, at best, extreme exaggerations (Anon., 1990).

Ramets of these salt tolerant clones have been tested in field trials on salt affected land. Some of the clones exhibited high growth rates, good form and tolerated very saline conditions. However, some other clones, even though selected under the same conditions in the greenhouse, did not survive well or died after growing for several years in the field, or had poor form, or were susceptible to insect damage (J.D. Morris, personal communication). This emphasizes the point that all genotypes, whether clones or seedlings, must be evaluated under field conditions over sufficient time before we can be confident of their superiority.

Some of the other species we are studying, including *E. nitens* and *E. globulus*, were selected for their high growth rate or pulp yield. These characteristics cannot be evaluated until the trees are at least 8 years old in temperate climates.

Simple methods of micropropagation

The scientific literature contains numerous references to procedures for the micropropagation of forest trees. In most cases these reports come from research laboratories where attention is of necessity paid to describing growth responses on complex and precisely formulated media; cultures are also grown under accurately controlled environmental conditions. However, once it has been shown that a particular species can be micropropagated then it is possible to simplify procedures in order to reduce the cost of commercial micropropagation (de Fossard and Bourne, 1977). We have developed simple commercial procedures for all of the species listed in Table 2.1. Some of the ways that we have used to simplify procedures and to reduce costs are discussed below.

Media formulations

It has been possible to micropropagate all of the above species on one, or at most two, basal mineral salt formulations. Eucalypts and acacias grow well on Murashige and Skoog salts while *Melaleuca* species generally grow best on Woody Plant Medium mineral salts (George *et al.*, 1987; Lloyd and McCown, 1981).

It has only been necessary to modify the hormone levels to a minor degree to achieve successful micropropagation of all of these species. Shoot multiplication occurs on media containing mineral salts, 2% sucrose, $1\,\mu$mol l^{-1} of BAP (benzylaminopurine) and $1\,\mu$mol l^{-1} of NAA (naphthalene acetic acid).

Root formation *in vitro* occurs on mineral salts, 2% sucrose and $10\,\mu$mol l^{-1} IBA (indole butyric acid).

No additions of organic compounds (vitamins, amino acids etc.) are essential to achieve micropropagation of any of these species. In some cases organic compounds are inhibitory to growth and root development.

Although the above media may not be optimal for a particular species, the fact that such a large number of species can be micropropagated on the same media represents a considerable saving for commercial laboratories.

Containers

Species have been cultured in a wide range of autoclavable containers (glass, polypropylene and polycarbonate) and pre-sterilized polystyrene Petri dishes. Stacks of Petri dishes or stacks of polypropylene disposable food containers are very cheap and enable large numbers of plants to be produced in a very small space. If containers are stacked on top of each other to save space then arranging fluorescent tubes vertically will provide more even illumination of the cultures than fluorescent tubes arranged horizontally along shelves.

Growth rooms

Cultures can be grown in simple growth rooms with imprecise temperature control. We have grown cultures in rooms with temperature fluctuations of at least ±5°C and in shaded greenhouses (about 80% shade). One successful commercial laboratory in Australia grows all its cultures in stacks of polypropylene food containers under natural light in shaded polythene greenhouses (L. Fumeaux, personal communication) (Fig. 2.2).

The saving in capital costs in such a laboratory compared with an air conditioned, artificially lit growth room is enormous.

Laboratory facilities

Expensive laboratory facilities are not essential for micropropagation. Mineral salts and hormone stock solutions can be purchased pre-mixed so avoiding the cost of expensive balances. In many cases standard mineral salt formulations used for potted plants or hydroponics are adequate. We have grown many species in Hoagland and Arnon's nutrient solution (George *et al.*, 1987) which we routinely use for potted plants. Good quality tap water can often be used instead of deionized or distilled water.

Large autoclaves and laminar flow transfer chambers are major capital costs in setting up a micropropagation laboratory. These expenses can be reduced by using alternatives such as pressure cookers instead of auto-claves. Laminar flow transfer chambers can be replaced with glove boxes or other 'still-air' transfer chambers. The commercial laboratory mentioned above manufactures its own transfer chambers out of glass aquariums with plastic sleeves to avoid contamination by the operator (Fig. 2.3). Sterility inside the aquarium is produced by spraying a chlorine solution (or other chemical sterilant) with a hand-held atomizer over the work area. We have run sterility tests (by exposing microbial plates) in the above chambers under operational conditions. Contamination was not detected even after leaving the plates exposed for several days.

Fig. 2.2. A commercial tissue culture laboratory where cultures are grown in stacks of polypropylene containers inside a plastic covered greenhouse.

Root formation *ex vitro* and in simple acclimatization facilities

Elimination of the necessity for root formation *in vitro* represents a saving in the most labour intensive part of the micropropagation process. Placing shoots on to a rooting medium usually involves selecting and cutting single shoots of a specified size. This procedure is much slower than cutting shoots for shoot multiplication medium where clumps of shoots of an irregular size can be used.

Many species can be rooted *ex vitro* (see Table 2.1); shoots are cut under non-sterile conditions (a simpler and cheaper option than cutting under sterile conditions) and placed directly into nursery trays containing a potting mixture. The shoots are then placed under intermittent mist in a shaded propagation bed for root formation and later acclimatization. This

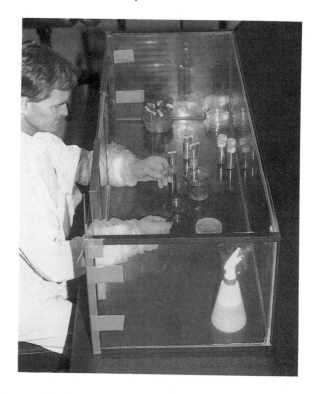

Fig. 2.3. An example of an aseptic transfer chamber manufactured from a glass aquarium.

procedure has worked particularly well for *E. camaldulensis*, *E. tereticornis* and some *Melaleuca* species.

However, mist propagation units can be expensive (especially if bottom heating is required) and not all species can readily root *ex vitro*. We have developed methods which simplify rooting *ex vitro* and, if necessary, rooting *in vitro*.

In the first method, shoots are placed on a sterile rooting medium *in vitro* and once root formation has occurred the lids on the containers are replaced with food wrapping plastics (many of these are permeable to water vapour) (Fig. 2.4) to allow acclimatization of the plants *in vitro* as water vapour is gradually lost through the plastic film.

It is often an advantage to use a potting mixture (such as sand, perlite, vermiculite or nursery plugs) as the solid support for the rooting medium rather than agar. Such physical supports reduce root damage and enable easier transfer of plants into nursery containers than removing rooted plants from agar (Fig. 2.5)

In the second method, plants are rooted *ex vitro* in 'miniature green-

Fig. 2.4. Rooted plants growing in vermiculite being acclimatized by covering the containers with plastic film permeable to water vapour.

houses' made from commercial plastic containers (Fig. 2.6). These containers are used in the take-away food industry or as packaging for food such as berries and sprouts. They are particularly useful for rooting *ex vitro* and in our experience they are much simpler, cheaper and produce faster growth than growing plants under intermittent mist.

If the containers do not have drainage holes these can be easily made in the base. The top of the container provides sufficient room for shoot growth and the top can be lifted gradually to allow the plants to acclimatize. These containers can be placed in the culture room or shaded greenhouse and sub-irrigated with water and nutrients when required. Plants can also be reliably transported in such containers.

The productivity per unit area must be taken into account when comparing micropropagation with the cost of producing cuttings from clone banks. Our calculations show that the productivity of shoots per unit area in a micropropagation laboratory is at least 400 times greater than that obtained from hedges grown to produce cuttings. Stacks of containers in a small micropropagation laboratory could easily have the same production capacity (as well as greater control) as several hectares of hedges in the field.

Fig. 2.5. Plants rooted *in vitro* in peat plugs ready for transfer to standard nursery trays for growing-on.

The main point of the above methods is not to provide yet another 'recipe' for the micropropagation of forest trees, but rather to suggest to nursery managers that there are numerous ways procedures can be simplified, steps eliminated, and production costs reduced. The exact techniques adopted will depend on local conditions and supply of materials.

Micropropagation need not be more expensive, laborious or demand much higher capital equipment than propagation by cuttings. Cuttings are now produced all over the world in facilities that range from sophisticated to very simple (Darus, 1991; Leakey, 1987; Simsiri, 1991; White 1988).

Long-term storage of clones *in vitro*

The ability to store clones in a disease-free state for long periods of time is a major advantage of micropropagation techniques. As clonal forestry increases in importance throughout the world, the ability to maintain disease-free foundation stock for future plantations and breeding programmes will be of major importance. The notable example of the use of tissue culture techniques for banana and plantain crops discussed in this

Fig. 2.6. Rooting plants *ex vitro* in 'miniature greenhouses' made from plastic food containers.

volume illustrates the importance of maintaining disease-free foundation stock.

Very few virus and other systemic diseases are known for forest trees (Büchen-Osmond *et al.*, 1988) but as our knowledge increases they are likely to become more noticed and important, just as has occurred with other domesticated crop plants.

Storage of plants *in vitro* also has other advantages.

1. It reduces the labour involved in regularly subculturing material. This is a major cost in running a commercial laboratory.
2. It means that a production facility can run at optimum capacity throughout the year with plants being stock-piled to meet seasonal demands e.g. for a restricted planting season.
3. It usually takes several years to evaluate the field performance of forest trees. Clones could be held under long-term storage conditions until field results and selections are made. If this is done, clones would have to be assessed for possible loss of vigour after long-term storage.

For all of the species listed in Table 2.1 we have tested three environments for the long-term conservation of clones as shoot culture *in vitro*:

1. 4°C, cool-white fluorescent lighting at 20 μmole m^{-2}s^{-1}, 8 h photoperiod;
2. 22°C $\pm$ 5°C, cool-white fluorescent lighting at 60 μmole m^{-2}s^{-1}, 16 h photoperiod;
3. 10°C, cool-white fluorescent lighting at 25 μmole m^{-2}s^{-1}, 16 h photoperiod.

Not all clones have been in all environments for the same period of time.

Of the 116 clones tested in both the 4°C and 22°C environments, 60% (70 clones) can easily survive for more than 6 months. Clones that can survive storage at 4°C can be stored for longer periods at 4°C than at 22°C.

We are currently testing 10°C as an environment where it may be possible to store tropical as well as temperate tree species. Ideally, we would like to store all clones in one environment.

Clones of some species have survived for long periods in the 4°C environment; several clones of *E. camaldulensis* and one clone of *E. grandis* for over 3 years, and several clones of *Melaleuca* and *E. globulus* subsp. *bicostata* for over 2 years.

Some clones of *E. camaldulensis* have survived as rooted plants *in vitro* in the 22°C environment for over 4 years without subculturing. The long-term survival ability of rooted plants compared with shoot cultures is also being tested at present.

Several problems have emerged during these experiments. The first problem is that there is considerable variation in survival between clones of a species under long-term storage. Some clones survive well while others deteriorate after only a few months. Deterioration is often associated with the appearance of slow-growing cryptic micro-organisms in culture. Slow-growing cryptic micro-organisms are a problem in many commercial laboratories. (Debergh and Maene, 1984). These organisms are not necessarily pathogenic to plants in the field but they represent a serious problem during the *in vitro* stages of production. Some of them probably represent contamination from human operators during subculturing, others may be endogenous.

We index cultures on microbial media and maintain high-health lines (as meristems or small shoot tips) of our most important species on a routine basis in order to reduce the problem of covert contamination.

Accurate recording and checking of data entries is essential for the maintenance of gene-banks for conservation and long-term storage of clones. A computerized database was developed in Canberra for the maintenance of records in a tissue culture laboratory and checking for any possible sources of error (Wolf and Hartney, 1986).

Micropropagation and inoculation with symbiotic micro-organisms

Since micropropagation techniques enable trees to be grown *in vitro* without competition from other organisms (axenic cultures) they are ideal tools to ensure establishment of a symbiotic association between selected trees and selected micro-organisms. This procedure is more reliable than alternative methods (Malajczuk and Hartney, 1986) and also allows automation of inoculation *in vitro* using alginate beads (Tommerup *et al.*, 1987).

Symbiotic associations between forest trees and soil micro-organisms have been found to be beneficial to tree growth and survival, especially under conditions of low soil fertility or various forms of environmental stress such as soil acidity or salinity (Marks and Kozlowski, 1973). Since many future plantings of forest trees will be on degraded land or land unsuitable for agriculture, symbiotic micro-organisms may have an important role to play in tree establishment and growth (Reddell and Warren, 1986).

On fertile sites, ectomycorrhizal fungi in association with the roots of eucalypts assist in the acquisition of phosphorus and growth of trees (Bougher *et al.*, 1990; Malajczuk *et al.*, in press). Atmospheric nitrogen fixation by *Frankia* species in association with *Casuarina*, and *Rhizobium* species in association with *Acacia*, plays an important part in the nitrogen nutrition of these species and can lead to the improvement of soil fertility on degraded sites (Graham *et al.*, 1988).

Genetic variation occurs in the micro-organism component as well as in the tree component of the above symbiotic associations. Different species and isolates of species of mycorrhizal fungi have been found to vary in their ability to produce symbiotic associations with a particular tree species and in their ability to produce large differences in the growth rate of trees, especially under conditions of low soil fertility (Malajczuk *et al.*, in press; Marks and Kozlowski, 1973). Mycorrhizal fungi have also been selected for adaptation to water stress (Coleman *et al.*, 1989), highly saline, alkaline and acid soils, and for the rehabilitation of mining sites (Gardner and Malajczuk, 1988).

Similarly, a large amount of genetic variation and specificity has been found in the ability of *Frankia* and *Rhizobium* species to establish nodules and fix nitrogen in association with their tree hosts (Midgley *et al.*, 1983; Turnbull, 1986a,b).

The first few years after planting is often the most critical time period in tree establishment. High mortality can occur at this time, especially on difficult sites. The fact that micropropagation techniques enable selected symbiotic micro-organisms to be established reliably *in vitro* may confer an advantage to inoculated trees during establishment.

Automation of micropropagation

Micropropagation has two inherent advantages over clonal propagation by cuttings. Firstly, the multiplication rates that can be achieved are usually much higher than for cuttings, and secondly, micropropagation allows year round production with control over most facets of production. With cuttings, on the other hand, seasonal peaks occur both in the production of cutting material from a clone bank and the rooting ability of cuttings (Denison and Quaile, 1987; Ikemori, 1984).

Year round production and precise control are usually prerequisites for the development of an automated system for any process. The other impetus to introduce automation is to reduce unit costs for large-scale operations.

The demand for forest trees is enormous compared with the demand for other plantation crops. For example, in the southeastern USA an average of more than 800 million seedlings per annum have been produced from seed orchard seed over the last 10 years (North Carolina State University–Industry Co-operative Tree Improvement Program, 1990). The clonal plantation programme of Aracruz in Brazil (based on eucalypt hybrids) probably represents the largest vegetative propagation programme of any crop in the world with about 15 million cuttings being produced in one year (Campinhos, 1984). In the future, as the clonal option for other forest trees becomes more attractive, or central micropropagation laboratories start to provide specialized services, then the demand for automated production systems will increase.

Automated production systems have been reported from several micropropagation laboratories. The automation of media preparation and sterilization is now routine or can be readily adapted from the food or pharmaceutical industries. Various liquid feeding systems (Aitken-Christie, 1986; Tisserat and Vandercock, 1985) and bioreactor devices (Takayama *et al.*, 1986) have been developed. However, the latter are not applicable to all crops and they do not overcome the most labour intensive and costly part of the micropropagation process: the selection, cutting and transfer of shoots on to fresh medium.

Commonwealth Industrial Gases (CIG) in Australia have developed an automated system which performs vision analysis of shoots, automatically cuts and transfers shoots from one container to another, or transfers shoots into nursery trays for growing-on (Johnson, 1989). This system has an operational speed several times faster than a human operator, can operate continuously with the minimum of operator supervision and has a production capacity in the order of millions of plants per annum.

The decision to automate micropropagation procedures must eventually be based on the relative costs of automated versus manual systems. We have illustrated above how many forest trees can be micropropagated as simply as cuttings. Automation will play an important role for those species where large scale industrial plantations are required.

Practical issues in clonal forestry

Two major problems are often raised in association with clonal forestry. The first problem is that only juvenile or 'rejuvenated' shoots have the ability to form roots. The second problem is that the planting of clones would reduce genetic diversity and leave our forests prone to disease and insect pest epidemics. We believe that both of these problems are more perceived than real.

There are substantive problems associated with clonal forestry. One example is that we tend to concentrate on a particular technique (e.g. micropropagation, cuttings, interspecific hybridization, genetic engineering) rather than look at the broad picture: the use of these techniques in tree improvement. This is particularly true with some of the claims being made for genetic engineering at present. These issues are discussed below.

Propagation from juvenile material and rejuvenation

The loss of rooting ability with age is common among woody perennials (Hackett, 1988). Shoots taken from older trees or from higher up the stem show a decline in rooting ability. Once trees have been evaluated in field trials, they will be several years old and shoots taken from the crowns will usually have lost their ability to form roots (Hartney, 1980). This is not a serious problem for many Australian genera since they retain the ability to form basal coppice shoots when trees are cut down or partially girdled. Cuttings taken from basal coppice shoots of eucalypts form the basis of the clonal propagation programme in Brazil (Brandão, 1984), South Africa (Denison and Quaile, 1987) and the Congo (Delwaulle, 1985; Souvannavong, p. 109 this volume).

Even for species that do not coppice readily (such as *E. nitens*) it is sometimes possible to obtain basal coppice shoots from young trees in some seasons, or by applying techniques such as partial girdling and exposing the stumps to full sunlight. Seasonal studies of coppicing ability for each species would be required in each location over different seasons.

If basal coppice shoots cannot be obtained then various 'rejuvenation' treatments could be tried. Grafting of small scions on to juvenile rootstocks, serial grafting, several subculturing cycles *in vitro*, and serial propagation by cuttings are some of the techniques used to stimulate adult cuttings to form roots (Franclet and Boulay, 1989; Hackett, 1988; Hartney, 1980).

The term 'rejuvenation' is used loosely in the literature. The ability to form roots may be regained using the above techniques. However, these ramets do not usually show complete rejuvenation equivalent to seedlings. They often quickly revert to adult leaf morphology, flower sooner and may grow more slowly than seedlings of the same age (Hackett, 1985; Pierik, 1990).

If none of these techniques is feasible and if vegetative propagation of elite clones is part of a tree improvement programme, then it is always possible to retain some members of a clone in a juvenile state. Seedlings can be maintained for years as small potted plants (by regular pruning) and still retain the ability to form roots. Selected families can be simply retained as seed, or clones can be stored *in vitro* until selections are made.

It is our opinion that rejuvenation of older trees has been perceived as a major problem and this has inhibited the introduction of vegetative propagation into tree improvement programmes. If it is desired to utilize vegetative propagation then some members of a clone or selected family should be retained in a juvenile state.

Since clones must be evaluated through their ramets it is essential to consider the method of vegetative propagation because of the important effects of age and method of propagation on growth. Physiological ageing is a real phenomenon with both positive and negative effects. Vegetative propagules from older shoots usually have more mature wood character- istics, less vigorous branching and better form; but they may also have a slower growth rate (McKeand, 1985; Martin, 1987; Sweet and Wells, 1974).

Also, it may not be possible to transfer the technology developed for one species to another. For example, the technique of producing cuttings of *E. grandis* hybrids from basal coppice shoots in Brazil was not completely satisfactory when applied to *E. globulus* subsp. *globulus* in Portugal. Cuttings from a hedged clone bank of *E. globulus* subsp. *globulus* were difficult to root, had leaves similar to older trees, flowered early, and grew more slowly than seedlings (K.G. Eldridge, personal communication).

Epidemics in clonal forests

Planting large areas with only one or a few genotypes may encourage development of epidemic populations of pathogens and pests by placing selection pressure on these organisms to overcome any inherent resistance in our forests. The breeding and selection of new genotypes able to resist diseases and pests is a major objective of most breeding programmes throughout the world.

Maintenance of genetic variation is the best guarantee that we will be able to respond to pests and diseases now and in the future. In this regard, tree breeders are probably in a better position than breeders of many other crop plants where populations of wild relatives have frequently been lost.

The conservation of genotypes *in situ, ex situ* and in gene banks should be a major international effort of tree breeding organizations (Burley, 1985; International Board for Plant Genetic Resources, 1984; Ledig, 1988).

In this light, clonal plantations do not necessarily represent a greater risk than plantations derived from seedlings. In clonal plantations we can maintain genetic variation over time by planting different clones in different years, genetic variation over the plantation area by planting mixtures of clones or planting different clones in separate mono-clonal blocks. In any case, many plantations grown from seedlings have been produced from a very restricted genetic base, in some cases with only a single tree as their genetic base (Eldridge, 1990).

Theoretical studies have suggested that the number of clones required to reduce the probability of epidemics need not be large (Libby, 1982; Lindgren *et al.*, 1989). In Europe, clonal plantations of poplar and spruce use very large numbers of clones in mixtures. No guarantee can be made that disease and insect pest epidemics will not occur in the future no matter how many clones we choose to plant. Serious epidemics of forest trees in recent years, for example Dutch elm disease, chestnut blight and pine wood nematode, have all occurred in seedling populations and devastated a vast array of genotypes.

Being able to utilize interspecific hybrids and progeny from controlled pollinations in clonal plantations creates a much wider range of genotypes available for selection than is possible within a single species. Selection of individual clones within a range of genotypes also means that selection for resistance can be much more intensive than selection of parents of seedlings.

Intensively managed plantations are usually much more profitable than less intensively managed plantations or natural forests (Dargavel and Cromer, 1979). Profits could be used to support a breeding and selection programme or to justify a chemical or biological control programme for pests and diseases. If a forest enterprise has low profitability, then insect and disease control may not be justified on economic grounds.

Genetic engineering

Considerable public and scientific interest has been focused in recent years on the prospects of transferring genes from other organisms into forest trees by genetic engineering. Unfortunately, some of the claims made for these techniques have been grossly exaggerated and it is prudent to put them into perspective.

Firstly, the ability to transfer genes between species (say herbicide tolerance from a bacterium into a forest tree) is to date restricted to the transfer of single genes, whereas many of our most valuable traits (for growth rate and disease resistance) are due to the interaction of many genes. Single-gene traits are also the easiest ones for disease organisms and pests to overcome by mutation.

Secondly, the production of genetic variation (by exploiting natural variation, crossing or genetic engineering) is only a small (and often the easiest and quickest) part of a tree improvement programme. New genotypes must then be evaluated in well-designed field trials, on a range of sites for a number of years. The time involved varies with location and species, but it is usually about one-half of rotation length, i.e. not less than 3–6 years even with fast growing species in tropical countries (Ikemori, 1984). With intensively bred crop plants like hybrid maize and wheat the release of new commercial varieties takes several years.

It is naive to assume that we can randomly insert single genes into an existing highly selected genotype and not obtain adverse effects resulting from interaction of the inserted gene with the rest of the genome, or suffer loss in growth or some other desirable characteristic. Plants do not function this way.

The above cautionary comments are not meant to be negative towards genetic engineering or other forms of biotechnology. There is no doubt that the ability to examine specific genes and their products and to map genomes will be of enormous importance to basic biology, and in understanding how organisms function. Rather, the comments are made as an appeal to take a broad view when we are determining priorities for tree breeding research. Do we know the right species, have we selected the right provenance, is vegetative propagation possible, can interspecific hybrids be useful for combining traits, how much gain will different methods provide and what will they cost? All of these questions, and many others, must be asked before we decide on the best method of tree improvement.

Other biotechnology techniques such as regeneration from disorganized tissue to obtain rejuvenation, embryo rescue of hybrids and somatic embryogenesis may have a more immediate role to play in tree improvement than genetic engineering. Tissue culture techniques play a central role in all aspects of biotechnology (McCown, 1985) and micropropagation is likely to be the method used to mass-propagate genetically engineered trees for plantations.

Conclusions

Micropropagation of Australian tree species has an important part to play in their domestication and as a part of tree improvement programmes.

Simple micropropagation techniques have been developed that do not demand expensive facilities or large amounts of capital equipment. Such techniques could be used to produce planting stock and to store clones *in vitro* for months.

Inoculation of trees with selected symbiotic organisms can also be readily incorporated into micropropagation procedures. Such symbiotic associations have assisted tree establishment on infertile sites.

Automation of micropropagation procedures has been developed and offers scope for large-scale clonal plantations in the future.

Acknowledgements

We gratefully acknowledge financial support from ACIAR (The Australian Centre for International Agricultural Research) for Mr V. Hartney to attend the symposium on the Rapid Propagation of Tree Species in Paris and to visit scientists involved in vegetative propagation of eucalypts in France and Morocco. We also thank Dr P. Kriedemann, Dr K. Eldridge and Mr R. Cromer for reviewing, Mr J. Dros and Mr M. Crowe for preparing the figures, and Mrs K. Munro and Mrs E. Morrow for typing the manuscript.

References

Aitken-Christie, J. (1986) Towards automated shoot production – nutrient regimes. In: *VI International Congress of Plant Tissue and Cell Culture, Abstracts.* International Association for Plant Tissue and Cell Culture, Minnesota, p. 453.

Anon. (1990) Super trees rehabilitate salt-damaged land and yield crop. *Australian Science and Technology Newsletter* 2, 1.

Aswathappa, N., Marcar, N.E. & Thomson, L.A.J. (1987) Screening of Australian tropical and subtropical tree and shrub species for salt tolerance. In: Rana, R.S. (ed.), *Afforestation of Salt Affected Soils* Vol. 2. Central Soil Salinity Research Institute, Karnal, India, pp. 1–15.

Boland, D.J., Brooker, M.I.H., Chippendale, G.M., Hall, N., Hyland, B.P.M., Johnston, R.D., Klening, D.A. & Turner, J.D. (1984) *Forest Trees of Australia* 4th edn. Nelson–CSIRO, Melbourne.

Bougher, N.L., Grove, T.S. & Malajczuk, N. (1990) Growth and phosphorus acquisition of karri (*Eucalyptus diversicolor* F. Muell.) seedlings inoculated with ectomycorrhizal fungi in relation to phosphorus supply. *New Phytologist* 114, 77–85.

Brandão, L.G. (1984) Presentation. In: *1. The New Eucalypt Forest.* The Marcus Wallenberg Foundation Symposium Proceedings 1984. The Marcus Wallenberg Foundation, Sweden, pp. 3–15.

Büchen-Osmond, C., Crabtree, K., Gibbs, A. & McLean G. (eds) (1988) *Viruses of Plants in Australia.* The Australian National University, Canberra.

Burley, J. (1985) *Global Needs and Problems of the Collection, Storage and Distribution of Multipurpose Tree Germplasm.* International Centre for Research in Agroforestry, Nairobi.

Campinhos, E., Jr. (1984) Presentation. In: *1. The New Eucalypt Forest.* The Marcus Wallenberg Foundation Symposium Proceedings 1984. The Marcus

Wallenberg Foundation, Sweden, pp. 21–7.

Coleman, M.D., Bledsoe, C.S. & Lopushinsky, W. (1989) Pure culture response of ectomycorrhizal fungi to imposed water stress. *Canadian Journal of Botany* 67, 29–39.

Dargavel, J.B. & Cromer, R.N. (1979) Pulpwood, money and energy. *Australian Forestry* 42, 200–6.

Darus, A.H. (1991) Multiplication of *Acacia mangium* by stem cuttings and tissue culture techniques. In: Turnbull, John W. (ed.), *Advances in Tropical Acacia Research*. Proceedings of a workshop held in Bangkok, Thailand, 11–15 February 1991. ACIAR Proceedings No. 35 (Australian Centre for International Agricultural Research), Canberra.

Debergh, P. & Maene, L. (1984) Pathological and physiological problems related to the *in vitro* culture of plants. *Parasitica* 40, 69–75.

de Fossard, R.A. & Bourne, R.A. (1977) Reducing tissue culture costs for commercial propagation. *Acta Horticulturae* 78, 37–44.

Delwaulle, J.C. (1985) Plantations clonales d'eucalyptus hybrides au Congo. *Revue Bois et Forêts des Tropiques*, no. 208, 37–42.

Denison, N.P. & Quaile, D.R. (1987) The applied clonal eucalypt programme in Mondi Forests. *South African Forestry Journal* 142, 60–7.

Eldridge, K.G. (1990) Conservation of forest genetic resources with particular reference to Eucalyptus species. *Commonwealth Forestry Review* 69, 45–53.

Franclet, A. & Boulay, M. (1989) Rejuvenation and clonal silviculture for *Eucalyptus* and forest species harvested through short rotation. In: Pereira, J.S. & Landsberg, J.J. (eds), *Biomass Production by Fast-Growing Trees*. Kluwer Academic Publishing, Dordrecht.

Gardner, J.H. & Malajczuk, N. (1988) Recolonization of rehabilitated bauxite mine sites in Western Australia by mycorrhizal fungi. *Forest Ecology and Management* 24, 27–42.

George, E.F., Puttock, D.J.M. & George, H.J. (1987) *Plant Culture Media*. Exegetics Limited, England.

Graham, P.H., Bale, J., Baker, D., Fried, M., Roskoski, J., Mackay, K.T. & Crasswell, E. (1988) The contribution of biological nitrogen fixation to plant production: an overview of the symposium and its implications. *Plant and Soil* 108, 1–6.

Hackett, W.P. (1985) Juvenility, maturation and rejuvenation in woody plants. *Horticultural Reviews* 7, 109–55.

Hackett, W.P. (1988) Donor plant maturation and adventitious root formation. In: Davis, T.D., Haissiq, B.E. & Sankhla, N. (eds), *Adventitious Root Formation in Cuttings*. Dioscorides Press, Portland, pp. 11–28.

Hartney, V.J. (1980) Vegetative propagation of the eucalypts. *Australian Forest Research* 10, 191–211.

Hartney, V.J. (1982) Tissue culture of *Eucalyptus*. *International Plant Propagator's Society, Combined Proceedings* 32, 98–109.

Hartney, V.J. & Kabay, E.D. (1984) From tissue culture to forest trees. *International Plant Propagator's Society, Combined Proceedings* 34, 93–9.

Ikemori, Y.K. (1984) Presentation. In: *1. The New Eucalypt Forest*. The Marcus Wallenberg Foundation Symposium Proceedings 1984. The Marcus Wallenberg Foundation, Sweden, pp. 16–20.

International Board for Plant Genetic Resources (1984) *The Potential for Using* in vitro *Techniques for Collecting Germplasm.* International Board for Plant Genetic Resources, Rome.

Johnson, B.J. (1989) Towards automation of crop plant tissue culture. *Australian Journal of Biotechnology* 3, 278–80.

Leakey, R.B. (1987) Clonal forestry in the tropics – a review of developments, strategies and opportunities. *Commonwealth Forestry Review* 66, 61–75.

Ledig, F.T. (1988) The conservation of diversity in forest trees. *Bioscience* 38, 471–9.

Libby, W.J. (1982) What is a safe number of clones per plantation? In: Heybrock, H.M., Stephan, B.R. & von Weissenberg, K. (eds), *Resistance to Diseases and Pests.* Pudoc. Wageningen, The Netherlands, pp. 342–60.

Lindgren, D., Libby, W.J. & Bondesson, F.L. (1989) Deployment to plantations of numbers of proportions of clones with special emphasis on maximizing gain at constant diversity. *Theoretical and Applied Genetics* 77, 825–31.

Lloyd, G. & McCown, B. (1981) Commercially-feasible micropropagation of mountain laurel, *Kalsmia latifolia*, by use of shoot tip culture. *International Plant Propagator's Society, Combined Proceedings* 30, 421–7.

McComb, J.A. & Bennett, I.J. (1986) Eucalyptus (*Eucalyptus* spp.) In: Bajaj, Y.P.S. (ed.), *Biotechnology in Agriculture and Forestry 1. Trees 1.* Springer Verlag, Berlin. pp. 340–62.

McCown, B.H. (1985) From gene manipulation to forest establishment: shoot cultures of woody plants can be a central tool. *Tappi Journal* 68, 116–19.

McKeand, S.E. (1985) Expression of mature characteristics by tissue culture plantlets derived from embryos of Loblolly pine. *Journal of the American Society for Horticultural Science* 110, 619–23.

Malajczuk, N., Grove, T.S., Tommerup, I.C., Bougher, N.L., Thomson, D.B., Dell pagated plantlets of *Eucalyptus camaldulensis* with ectomycorrhizal fungi and comparison with seedling inoculation using inoculum contained in a peat/vermiculite carrier. *Australian Forest Research* 16, 199–206.

Malajczuk, N., Grove, T.S., Tommerup, I.C., Bougher, N.L., Thomson, D.B., Dell, B. & Kuek, C. Ectomycorrhizas. In: Dart, P. & Dawson, J.O. (eds), *Microorganisms that Promote Plant Productivity.* Martinus Nijhoff, Amsterdam, (in press).

Marcar, N.E. (1989) Salt tolerance of frost-resistant eucalypts. *New Forests* 3, 141–9.

Marks, G.C. & Kozlowski, T.T. (eds) (1973) *Ectomycorrhizae: Their Ecology and Physiology.* Academic Press, New York.

Martin, B. (1987) 'Amelioration genetique des Eucalyptus tropicaux. Contribution majeure a la foresterie clonale.' Unpublished DSc Thesis, de l'Universite de Paris XI.

Matheson, A.C. & Lindgren, D. (1985) Gains from the clonal and clonal-seed orchard options compared for tree breeding programs. *Theoretical and Applied Genetics* 71, 242–9.

Midgley, S.J., Turnbull, J.W. & Johnston, R.D. (1983) *Casuarina Ecology, Management and Utilization.* Proceedings of an international workshop, Canberra, Australia, 17–21 August 1981. CSIRO, Canberra.

North Carolina State University – Industry Cooperative Tree Improvement Program

(1990) *Thirty-Fourth Annual Report.* College of Forest Resources, N.C. State University, Raleigh, North Carolina.

Pierik, R.L.M. (1990) Rejuvenation and micropropagation. *Newsletter, International Association for Plant Tissue Culture* No. 62, 11–21.

Reddell, P. & Warren, R. (1986) Inoculation of acacias with mycorrhizal fungi: potential benefits. In: *Australian Acacias in Developing Countries.* ACIAR Proceedings No. 16. ACIAR, Canberra, pp. 50–3.

Simsiri, A. (1991) Vegetative propagation of *Acacia auriculiformis.* In: Turnbull, John W. (ed.), *Advances in Tropical Acacia Research.* Proceedings of a workshop held in Bangkok, Thailand, 11–15 February 1991. ACIAR Proceedings No. 35 ACIAR, Canberra.

Sweet, G.B. & Wells, L.G. (1974) Comparison of the growth of vegetative propagules and seedlings of *Pinus radiata. New Zealand Journal of Forestry Science* 4, 399–409.

Takayama, S., Arima, Y. & Akita, M. (1986) Mass propagation of plants by fermenter culture techniques. In: *VI International Congress of Plant Tissue and Cell Culture, Abstracts.* International Association for Plant Tissue and Cell Culture, Minnesota, p. 449.

Tisserat, B. & Vandercock, C.E. (1985) Development of an automated plant culture system. *Plant Cell, Tissue and Organ Culture* 5, 107–17.

Tommerup, I.C., Kuek, C. & Malajczuk, N. (1987) Ectomycorrhizal inoculum production and utilization in Australia. In: *Mycorrhizae in the Next Decade – Practical Applications and Research Priorities.* Institute of Food and Agricultural Sciences, Florida.

Turnbull, C.R.A., Beadle, C.L., Bird, T. & McLeod, D. (1988) Volume production in intensively-managed plantations. *Appita* 41, 447–50.

Turnbull, J.W. (ed.) (1986a) *Australian Acacias in Developing Countries.* ACIAR Proceedings No. 16. ACIAR, Canberra.

Turnbull, J.W. (ed.) (1986b) *Multipurpose Australian Trees and Shrubs – Lesser known species for fuelwood and agroforestry.* ACIAR, Canberra.

White, K.J. (1988) *Applied Tree Breeding, Strategies, Practices, Programmes in Bhabar Terai of Central Nepal.* Manual No. 5. Sagarnath Forest Development Project, Ministry of Forests, Nepal.

Wolf, L.J. & Hartney, V.J. (1986) Computer system to assist with management of a tissue culture laboratory. *New Zealand Journal of Forestry Science* 16, 392–402.

Zobel, B.J., Wan Wyk, G. & Stahl, P. (1987) *Growing Exotic Forests.* John Wiley and Sons, New York.

Chapter 3

Rapid Clonal Propagation, Storage and Exchange of *Musa* spp.

Jos Schoofs

*INIBAP Transit Center, Catholic University of
Leuven, Belgium*

The main objectives of the International Network for the Improvement of Banana and Plantain (INIBAP) which was established in 1984, are to initiate, encourage, support and co-ordinate research carried out in different countries to ensure an adequate production of locally grown bananas and plantains. One of the Network's roles is to offer the International Community all sources of *Musa* germplasm (natural cultivars and improved hybrids).

Since the early 1980s, practical tissue culture techniques for rapid clonal propagation of *Musa* have been worked out in many laboratories. One is located at the Catholic University of Leuven (K.U.Leuven), Belgium. One of the reasons why INIBAP selected this laboratory as a centre for the international transfer and exchange of banana and plantain germplasm was that it is situated in a non-banana producing country where import regulations allow a smooth movement of germplasm. In this chapter the many advantages of *in vitro* propagation of bananas and plantains over traditional propagation are reviewed.

Rapid clonal propagation

Plant material

One of the most crucial concerns of *in vitro* culture is to retain cultivar characteristics because somaclonal variation, which is genetic variation among plants regenerated from tissue cultures of a single parental clone, appears to be a potential hindrance to the *in vitro* propagation, conservation and exchange of germplasm. Shoot tip culture, however, is generally considered as an *in vitro* culture system with low risk of genetic instability because of the large, organized structure of the explants. Notwithstanding this consideration, several reports on the incidence of off-types among micropropagated bananas and plantains exist (Hwang, 1986; Hwang and Ko, 1987; Krikorian, 1989; Müller and Sandoval, 1987; Pool and Irizarry, 1987; Ramcharan *et al.*, 1987; Reuveni *et al.*, 1985; Smith, 1988; Stover, 1987; Vuylsteke *et al.*, 1988). The bulk of these off-types, however, parallel naturally occurring variability and should thus not be considered as being a specific consequence of tissue culture. Moreover, many of the off-types can already be eliminated during the early developmental stages of young plants in the nursery (Vuylsteke *et al.*, 1988).

Shoot tips can be obtained from all apex-containing plant parts: the parental pseudostem, its suckers, peepers, lateral buds and eyes. Even the apex of the terminal inflorescence and the young flower buds are capable of reverting to vegetative meristems *in vitro*. There are no differences in growth response or explant survival in culture among explants from these sources, and neither the physiological age of the shoot tip nor the season in which the explants are obtained influence explant behaviour or performance in culture. However, buds and small sword suckers are the preferred source material due to their greater ease of handling. Plant material should only be collected from flowering source plants to ascertain trueness-to-type. Source plants should also be disease-free and growing vigorously. Special pretreatments of the mother plants, however, are not required.

Culture medium

The culture medium that is used at the INIBAP Transit Centre (see Table 3.1) is based on the Murashige and Skoog (1962) mineral salt mixture, which is very suitable for banana and plantain shoot tip culture and is most widely used.

1. The mineral salts are the same as in the Murashige and Skoog nutrient medium, except the concentration of phosphate is doubled.

2. Of the vitamins only thiamine seems to be necessary (Krikorian and Cronauer, 1984). The amino acid glycine is not essential but provides a source of nitrogen that is immediately available to cultured tissues (George and Sherrington, 1984).

3. Ascorbic acid or vitamin C is added as antioxidant in order to reduce blackening of shoot tips (Vuylsteke and De Langhe, 1985). Tissue blackening is caused by the oxidation of phenolic compounds that, upon wounding, exude into the medium where they accumulate and form a blackened area around the explant hindering nutrient uptake and causing growth inhibition. Blackening is particularly intense in newly initiated cultures and usually decreases over time.

Table 3.1. Composition of the modified Murashige and Skoog nutrient medium, used for tissue culture of bananas and plantains at the INIBAP Transit Centre at K.U. Leuven.

		$mg\,l^{-1}$	$\mu mol\,l^{-1}$
Macronutrients	NH_4NO_3	1650	20.63
	KNO_3	1900	18.81
	$CaCl_2.\,2H_2O$	440	2.99
	$MgSO_4.\,7H_2O$	370	1.50
	KH_2HPO_4	400	1.75
Micronutrients	H_3BO_3	6.18	100
	$MnSO_4.\,H_2O$	16.90	100
	$ZnSO_4.\,7H_2O$	8.60	30
	KI	0.83	5
	$Na_2MoO_4.\,2H_2O$	0.24	1
	$CoCl_2.\,6H_2O$	0.024	0.1
	$CuSO_4.\,5H_2O$	0.025	0.1
Iron	$FeSO_4.\,7H_2O$	27.80	100
	$Na_2.\,EDTA.\,2H_2O$	37.22	100
Vitamins	Glycine	2.0	26.64
	Thiamine hydrochloride	0.1	0.30
	Nicotinic acid	0.5	4.06
	Pyridoxine hydrochloride	0.5	2.43
Antioxidant	Ascorbic acid	10.0	56.78
Carbon source	Sucrose	30 000	87 642
Gelling agent	Gelrite[R]	2000	–
Growth regulators	N^6-benzylaminopurine	0.225*	1*
	Indole-3-acetic acid	0.175	1

*for multiplication and storage

4. Sucrose is the preferred carbon source. Sucrose of analytical grade, however, is expensive and is often replaced by refined grocery sugar.

5. An important physical aspect of the culture medium is its solid or liquid state. Although liquid cultures have a slightly increased growth rate and shoot proliferation, semi-solid cultures are preferred because they do not require expensive equipment, occupy less space in the culture rooms and give fewer problems when transferring plants to the external environment. Agar is widely used to solidify media, but it is expensive, contains contaminating compounds and the composition varies. Therefore, it is replaced as gelling agent by Gelrite[R], which not only gives up to 20% better results, but also forms clear gels that make it easier to detect contaminations by bacteria or fungi.

6. The pH is usually kept at 5.8 just before autoclaving. During auto-claving it drops 0.5–1.0 units, which sometimes affects the gelling effi-ciency of Gelrite^R or agar. The pH also changes during culture with acidity sometimes increasing to pH 4.5 during 1 month of culture, indicating the need for timely transplantation to fresh medium (Vuyl-steke and De Langhe, 1985).

Culture environment

For experimental purposes or for storage at the INIBAP Transit Centre at K.U.Leuven shoot tip cultures are established in 15 cm high test tubes of 2.5 cm diameter. The test tubes are filled with 20 ml of nutrient medium and closed with autoclavable plastic caps. For mass production large jars in glass or plastic or Petri dishes are more convenient as each test tube can hold only one explant.

Tissue cultures are incubated in rooms with controlled temperature and light regimes. Temperature should be maintained between 20°C and 35°C. Optimum incubation temperature is 30 ± 2°C. Artificial lighting is provided by cool-white, fluorescent tubes and cultures are maintained under continuous light of 5000 lx.

Culture methodology

The *in vitro* production of plants proceeds through a sequence of three major steps: the initiation of an aseptic culture; the multiplication of propagules; and the regeneration of plants for transfer to soil.

Initiation

Shoot tips are harvested from plants growing in the natural environment and are thus contaminated with many micro-organisms. Hence they must be surface sterilized before explants are transferred to culture medium. Shoot apices of bananas and plantains, however, are covered by many tightly overlapping leaf initials, which form a natural protection against surface contaminants and make sterilization easy.

Cubes of 2–8 cm^3 of healthy tissue containing the shoot tip are excised and washed in 70% ethanol for 30–60 s, followed by a 20 min treatment in a 5-times diluted commercial laundry bleach solution containing 10% w/v NaOCl. Traces of the disinfectant are removed by three rinses with sterile, distilled water lasting 10, 10 and 5 min.

Using a dissecting microscope excessive tissue from the sterile blocks of tissue is removed by trimming away the outer leaf initials and by reducing corm tissue to a thin slice. The explants thus obtained are cone-shaped and comprise a number of leaf primordia enclosing the meristem. They are about 8–10 mm high and have a circular sectional plane of 4–6 mm at their base. Explants that are smaller need too much time before growth resumes while bigger ones suffer from excessive blackening of the corm tissue. To reduce the risk of losing all explants due to accidental contamination and to speed up multiplication the shoot tips are often cut longitudinally in two halves.

Freshly excised explants are inoculated on a multiplication medium containing $10 \, \mu$moll^{-1} of cytokinin 6-benzyladenine. Other cytokinins, like kinetin or zeatin, give inferior results for the induction of shoot or bud proliferation *in vitro* (Krikorian and Cronauer, 1984; Vuylsteke and De Langhe, 1985). The medium also contains $1 \, \mu$moll^{-1} of the auxin indole-3-acetic acid, which can be replaced by 1-naphthalene acetic acid (NAA) or indole-3-butyric acid (IBA) (Vuylsteke, 1989).

Multiplication

The shoot tips turn green within 1 week and appear basally swollen after 2–3 weeks. Underneath the outermost leaves, a number of tiny meristems can be seen in a circular arrangement. A first subculture can be executed after 4–5 weeks of growth by subdividing the clusters of buds into smaller pieces of tissue. There is controversy over whether the newly formed lateral meristems that arise directly from the explant originate from axillary buds or as adventitious growths. This is not surprising, if one considers the differing views that already exist on the *in vivo* situation (Barker and Steward, 1962). Transfer to fresh nutrient medium takes place every 4–6 weeks. Newly initiated cultures often suffer from excessive blackening and need a transfer to fresh medium once a week during the first 1–2 months of culture.

All meristems used to initiate *in vitro* cultures of a particular plant receive a separate code number. Once cultures originating from one of these meristems are well established *in vitro*, all cultures originating from other meristems of the same plant are discarded. This procedure ensures that each accession at the INIBAP Transit Centre originates from a single meristem and guarantees 100% homogeneity of the cultures.

Not all bananas and plantains show the same form and rate of proliferation on the same culture medium. Some accessions do not multiply at all: they grow as a single plantlet without forming any new shoots or meristems. Other cultivars, on the contrary, multiply so fast that they do not even produce shoots or leaves, but only develop bulbil-like structures that bear numerous minute meristems. Many types of intermediate forms, containing both shoots and bulbil-like structures in different proportions, exist. Variation of the cytokinin content of the culture medium can make a particular accession change its multiplication rate and form, from one intermediate form to another or even from one extreme form to the other. Different forms of proliferation thus seem to be nothing more than variations of a single phenomenon that can gradually change under varying concentrations of cytokinin in the culture medium. Cytokinins reduce or suppress the dominance of the main meristem over other meristems in its vicinity. The concentration of cytokinin in the proliferation medium that is used at the INIBAP Transit Centre at K.U.Leuven is sufficiently high to suppress this apical dominance completely, or at least to a certain degree, in most of the banana and plantain cultivars as well as in most of the wild species. Bud formation can also be achieved by fragmenting the shoot apex. Apical dominance continues in only one of the fragments preventing proliferation, but in the others multiple bud or shoot formation will take place as they are mechanically separated from the main meristem.

Through manipulation of the cytokinin content of the nutrient medium a proliferation rate of 20–50 new meristems from a single meristem every 4–6 weeks can be obtained easily, which is already more than double the yearly yield of shoots of a plant in field conditions and this allows the production of millions of new plants a year.

Regeneration

Propagules obtained during the multiplication stage are very small and not yet capable of survival in soil. Elongation of buds into shoots, is easily accomplished by transferring individual shoots or clusters of small buds to regeneration medium. This differs from the proliferation medium in cytokinin content, which is reduced ten times to $1\,\mu\text{mol}\,l^{-1}$. Rooting can be induced on a medium with macronutrients at half concentration, a sugar content reduced to $10\,\text{g}\,l^{-1}$ sucrose and $1\,\mu\text{mol}\,l^{-1}$ of the stronger auxin IBA. On this medium, roots usually appear within a week.

Rooted shoots can be left for 3–4 weeks on the regeneration medium and then transferred directly to the external environment or they can undergo a hardening period before transplantation to soil if this should be necessary to achieve survival rates exceeding 90%. Hardening takes place in large, 250 ml jars containing a culture medium with the concentration of macronutrients reduced to half and no growth regulators. In such jars vigorous shoots, 8–10 cm tall with many long and well ramified roots, develop within 1 month.

Micropropagated plants are relatively delicate because they have been grown in high humidity and low light intensity and may lose water rapidly upon transfer to natural conditions. For successful *in vivo* establishment of *in vitro* produced plants a nursery area or glasshouse with some facilities for humidity control and partial shading is required. Plants are routinely potted in a 1:1 peat:soil mixture. A sterile soil or soil mixture is not essential for successful plant establishment, but is recommended in order to maintain the freedom from disease. Soil moisture is a critical factor in successful plant establishment and the line between too much and too little is very narrow. Four to six weeks after transplantation a dilute solution of fertilizer containing urea (0.1–0.5 g/100 ml/plant) and potash (0.2–1.0 g KCl/100 ml/plant) can be applied (Vuylsteke, 1989).

Disease status

Fungal and bacterial diseases are unlikely to occur on plants obtained through tissue culture. If fungi or bacteria survive the preceding disinfection procedures, they will become self-evident by contamination of the culture medium. All culture vessels showing the slightest trace of contamination are discarded automatically.

Fast dividing tissue in the vicinity of the meristem is often considered as free of viruses. A complete pathogen-free status of the cultures, however, cannot always be guaranteed as virus diseases may pass undetected through *in vitro* culture. Therefore, serological and molecular detection techniques are needed to guarantee the disease-free status of the cultures. These techniques exist at present for the cucumber mosaic virus (CMV) but not for the banana bunchy top virus (BBTV) and other viruses.

Conservation of germplasm

Field collections of vegetatively propagated crops are labour-intensive, occupy a great amount of space, are subject to disease and adverse weather conditions and the plants can only be multiplied very slowly when required. Storage *in vitro* under the above described growth conditions overcomes many of these problems but still needs a transfer of the cultures to fresh nutrient medium every 4–6 weeks. As this procedure would be too labour-intensive plants are kept under reduced growth conditions. These can be provided in two ways: (1) modification of the nutrient medium or (2) modification of the physical environment in which cultures are incubated. As the former option seemed to be too complicated and too time-consuming because of the need for different culture media for multiplication and storage, the latter was preferred at the INIBAP Transit Centre. Light conditions are reduced to about 2000 lx and 12 h, while temperature is kept at $15 \pm 2°C$.

Experiments at the INIBAP Transit Centre showed that cultures can be kept easily for 18 months under these conditions without transfer to fresh medium, but for absolute safety this period has been reduced to 12 months. Each accession is represented by a batch of 24 test tubes and cultures are checked every 4 weeks for contamination or necrosis.

Unsuitable cultures are discarded and when only 16 tubes of a particular accession are left, they are moved to normal growth conditions for multiplication on fresh nutrient medium.

New techniques like cryopreservation open completely new horizons of germplasm storage. At the INIBAP Transit Centre experiments showed that cell suspension cultures and clusters of cells can survive storage in liquid nitrogen after a pretreatment with a cryoprotectant (DMSO, glycerol, proline, sucrose). Viability was estimated through fluorescence with fluorescine diacetate (FDA) and confirmed with regrowth tests on semi-solid medium through somatic embryogenesis (Panis, personal communication, 1989).

Exchange of germplasm

The transfer from one country or continent to another of corms, suckers or other vegetative propagation material involves the risk of inadvertently spreading economically important pests and diseases. Therefore, many countries have adopted a very conservative attitude towards plant introduction in vegetative form and germplasm movement is often hindered by lengthy post-entry quarantine procedures. Not only the danger of spreading pests and diseases, but also their bulkiness, preclude the use of corms or suckers as an ideal means of transfer of germplasm. Tissue cultures can overcome many of these disadvantages.

The Technical Guidelines for the Safe Movement of *Musa* Germplasm, resulting from a combined effort of FAO/IBPGR and INIBAP, provide a number of recommendations that should be followed when transferring banana or plantain germplasm (Frison and Putter, 1989).

1. All germplasm should be transferred from one banana producing country to another as *in vitro* cultures.
2. Shoot tips or vegetatively propagated material should be sent to a transit centre in a non-banana producing country where the material should be subjected to shoot tip culture.
3. Multiplication *in vitro* should result in the production of replicates from a single original shoot tip, which will be used for indexing, distribution and storage.
4. For the movement of cultures, neither antibiotics nor charcoal should be added to the medium so as to facilitate detection of contamination by fungi or bacteria.
5. Four *in vitro* plantlets should be sent from the transit centre to one or two indexing centres where they will be indexed for virus diseases.
6. As soon as the indexing centres report negative results, the transit centre is in a position to provide *in vitro* duplicates to the region where the accession has been requested.

7. In the region of destination, the material should be multiplied in field conditions and remain under observation for a period of a year, in agreement with the local quarantine authorities.

For successful exchange of germplasm, culture vessels are very important. They should be unbreakable, airtight, transparent and not too bulky. Small 40 ml containers (55 × 20 mm) in rather thick, transparent plastic and with a screwcap closure have already proved to be very suitable in many shipments. They are filled with approximately 15 ml of culture medium solidified with 10 gl^{-1} agar. Gelrite® proved to be less suitable as it starts to liquify after only a week of culture, even when used in high concentrations.

Cultures are kept in these containers under normal growth conditions for at least seven days before shipment to verify that no accidental contamination took place during manipulation as this would inevitably mean discard of the cultures in quarantine. The separately packed containers are then placed in cartons lined with shock-absorbent material and shipped by courier with all necessary documents.

Up till now the INIBAP *Musa* Germplasm Transit Centre at K.U.Leuven has imported over 500 accessions from 30 different locations and nearly 1000 have been exported to around 25 destinations (see Table 3.2).

Whatever the field of research in which banana and plantain germplasm is used, information and documentation are essential resources for scientists. At present, only limited passport data of the accessions of most collections are available and information on agronomic value and characteristics is very scarce. As many accessions exist under different names in different regions throughout the world it is essential to collect all available passport data in order to eliminate duplicates as much as possible, since this would reduce the workload significantly.

Genetic improvement

Tissue culture has many indirect advantages for genetic improvement of banana and plantain by providing a safe means for exchange of valuable germplasm and a means for rapid multiplication of improved plant material.

Direct advantages, however, also exist. Somaclonal variation, for example dwarfism, can be of some agronomic importance. Besides these more accidental improvements, tissue culture can be of help in direct genetic improvement through protoplast fusion. At the INIBAP Transit Centre preliminary experiments showed that it is possible to establish protoplast cultures of *Musa* spp. and that it is possible to regain plants from

Table 3.2. *Musa* germplasm exchange activities at the INIBAP Transit Centre at K.U. Leuven (till 30 October, 1989).

A. *Number of imported accessions*

Region	1976–1983	1984	1985	1986	1987	1988	1989	Total
Latin America and Caribbean		2	2	5	53	66	12	140
West and Central Africa	5	15	23	51	35	41		170
East Africa				7	54			61
Asia and Pacific	6			5	6	21	70	108
Europe	5		2	1	30	22	19	79
Total for year	16	17	27	69	178	150	101	558

B. *Number of exported accessions*

Region	1985	1986	1987	1988	1989	Total
Latin America and Caribbean	20		13	127	87	227
West and Central Africa		27	154	105	109	415
East Africa				105	69	174
Asia and Pacific	12		3	5	33	53
Europe			10	14	29	53
Total for year	32	27	180	356	327	922

cell suspension cultures through somatic embryogenesis. These techniques open completely new horizons for genetic improvement of bananas and plantains.

References

Barker, W.G. and Steward, F.C. (1962) Growth and development of the banana plant. 1. The growing regions of the vegetative shoot. *Annals of Botany* 26, 389–411.

Frison, E.A. & Putter, C.A.J. (1989) *FAO/IBPGR Technical Guidelines for the Safe Movement of* Musa *Germplasm.* Food and Agriculture Organization of the United Nations/International Board for Plant Genetic Resources, Rome.

George, E.F. & Sherrington, P.D. (1984) *Plant Propagation by Tissue Culture.* Exegetics Ltd, Basingstoke, U.K.

Hwang, S.C. (1986) Variation in banana plants propagated through tissue culture. *Journal of the Chinese Society of Horticultural Science* 32, 117–125.

Hwang, S.C. & Ko, W.H. (1987) Somaclonal variation of bananas and screening for resistance to Fusarium wilt. In: Persley, G.J. & De Langhe, E.A. (eds), *Banana and Plantain Breeding Strategies. Proceedings International Workshop.* Cairns, Australia, 13–17 Oct., 1986. ACIAR Proceedings No. 21, 151–6.

Krikorian, A.D. & Cronauer, S.S. (1984) Banana. In: Sharp, W.R., Evans, D.A., Ammirato, P.V. & Yamada, Y. (eds), *Handbook of Plant Cell Culture*, Vol. 2. Macmillan, New York, pp. 327–48.

Krikorian, A.D. (1989) *In vitro* culture of bananas and plantains: background, update and call for information. *Tropical Agriculture (Trinidad)* 66, 194–200.

Müller, L.E. & Sandoval, J. (1987) The problem of somaclonal variation in *Musa* spp. In: *Proceedings 2nd Annual Conference of IBPNet*, Bangkok, Thailand, 11–16 Jan., 1986.

Murashige, T. & Skoog, F. (1962) A revised medium for rapid growth and bio-assays with tobacco tissue cultures. *Physiologica Plantarum* 15, 473–97.

Pool, D.J. & Irizarry, H. (1987) Off-type banana plants observed in a commercial planting of Grand Nain propagated using the *in vitro* culture technique. In: Galindo, J.J. & Jaramillo, R. (eds), Proceedings 7th ACORBAT Meeting, San José, Costa Rica, 23–27 Sept., 1985. *CATIE Technical Bulletin* 121, 99–102.

Ramcharan, C., Gonzalez, A. & Knausenberger, W.I. (1987) Performance of plantains produced from tissue-cultured plantlets in St Croix, U.S. Virgin Islands, *Proceedings 3rd IARPB Meeting*, Abidjan, Côte d'Ivoire, 27–31 May, 1985, pp. 36–9.

Reuveni, O., Israeli, Y., Degani, H. & Eshdat, Y. (1985) Genetic variability in banana plants multiplied via *in vitro* techniques. *Research Report AGPG*, IBPGR/85/216.

Smith, M.K. (1988) A review of factors influencing the genetic stability of micro-propagated bananas. *Fruits* 43, 219–23.

Stover, R.H. (1987) Somaclonal variation in Grand Naine and Saba bananas in the nursery and field. In: Persley, G.J. & De Langhe, E.A. (eds), *Banana and Plantain Breeding Strategies. Proceedings International Workshop.* Cairns,

Australia, 13–17 Oct., 1986. ACIAR Proceedings No. 21, 136–9.

Vuylsteke, D.R. (1989) *Shoot-tip Culture for the Propagation, Conservation and Exchange of* Musa *germplasm.* International Board for Plant Genetic Resources, Rome.

Vuylsteke, D. & De Langhe, E.A. (1985) Feasibility of *in vitro* propagation of bananas and plantains. *Tropical Agriculture (Trinidad)* 62, 323–8.

Vuylsteke, D., Swennen, R., Wilson, G.F. & De Langhe, E. (1988) Phenotypic variation among *in vitro* propagated plantain (*Musa* sp. cultivar AAB). *Scientia Horticulturae* 36, 79–88.

Chapter 4
Bamboo Propagation Through Conventional and *In Vitro* Techniques

I.V. Ramanuja Rao, I. Usha Rao and Farah Najam Roohi

Department of Botany, University of Delhi, Delhi
110007, India

Bamboos represent one of the world's great natural and renewable resources. They provide a versatile multipurpose forest product which plays a vital role in our domestic economy. The utility of bamboos stems from their strong, straight, smooth, light and hard nature. They are easy to exploit, transport, split, cut and fashion. Further, they are abundant, fast-growing and attain maturity in a relatively short period.

These giant arborescent grasses have woody culms (erect stems) arising from underground rhizomes. The culms which can be up to 40 m in height and 30 cm in diameter, reach their full height within just 4 months. The bamboos thus constitute one of the most rapidly growing components of a flora. An important characteristic of the plant is that it spreads laterally by outward growth of the rhizome. Bamboos, therefore, quickly colonize the land. Tufts of strong roots are produced at each node along the rhizome. These masses of roots together form an interlocked, mat-like structure. This feature together with the fact that the rhizome generally grows in the top 50 cm of the soil, serves to prevent erosion of precious top soil. The woody material has bundles which remain separate in the long internodes, giving bamboo an easy splitting characteristic that makes it so amenable to handicrafts. They also give it great flexibility. Bamboos are endowed with yet another feature: long fibre length, which makes bamboo pulp eminently suited to paper making. It also contains a relatively low proportion of lignin. In the tropics, therefore, bamboos are perhaps the best alternative to softwoods.

Bamboos are put to a wide variety of uses (Table 4.1) and have been intimately associated with the civilization of man from ancient times. In modern industrial society, characterized by plastics and steel, bamboos are still relevant and contribute much to human well-being.

Table 4.1. Uses of bamboo.

Agricultural implements	Extract for culture of pathogenic bacteria	Pan trays
As an aphrodisiac	Eyeliner	Paper
Bamboo shoots	Fencing	Peans, traps
Bamboo charcoal	Fibre-reinforced plastic	Pea sticks
Barbecue skewers	Fishing implements, floats	Plaited shoes
Barrels for toy cannon	Fishing rods	Plywood
Baskets	Floats for timber, rafts	Polo mallets
Boards	Fodder	Prevention of soil erosion – river-banks, etc.
Bows and arrows	Fuel	Punting poles
Boat roof	Furniture	Rickshaw hoods
Brush to make line images	Hats	Scaffoldings
Carbonized filament as a light giving element	Haystack stabilizers	Seed food
Cart yokes	Hedges	Shutters
Cart sheds, roofs, stakes, Boat masts	Hookah pipes	Sledges
Charcoal	Horticulture	Thatching and roofing
Cheroot wrapper	Houses	Tool handles
Chicks for doors and windows	Jaundice treatment	Toys
Construction	Joints for cooking glutinous rice	Trays for silk worms
Containers to administer medicine to animals	Ladders	Trellises
Cooling tonic	Lance staves	Umbrella, umbrella handles
Cooking utensils	Liquid diesel fuel	Walling
Cordage	Loading vessels	Walking sticks
Country tiles	Mats	Water and milk vessels
Cups, containers	Match boxes	Water pipes
Cradles	Musical instruments flutes, etc.	
Cremation – biers	Ornaments	

Conventional propagation methods

Conventionally, bamboos are propagated through seeds, or by vegetative means. If the seeds are good, regeneration is normally easy, since their small size makes them easily transportable. The difficulty with seeds is their low viability, poor storage characteristics and inborne microbial infestation. In addition, seed availability is uncertain due to the long vegetative habit of bamboos and depends on sporadic flowering. (In species such as *Dendrocalamus hamiltonii* where flowering takes place both sporadically and at long intervals, very little, if any fertile seed is formed. Vegetative methods are, therefore, the most usual way of propagation.)

Vegetatively, bamboos are propagated through divisions, macro-proliferation, rhizomes, offsets, layering, marcotting, culm and branch cuttings. For successful establishment and growth, a bamboo propagule must possess three structures: a well-developed root system, a rhizome and a shoot.

Divisions

The dwarf bamboos are usually propagated by divisions. The culms, in small clumps of two or three with attached rhizomes, are transplanted immediately before the growing season commences, with as much soil as possible around the rhizomes (Troup, 1921).

Macro-proliferation

Since bamboo seedlings possess inherent proliferating capacity and ability to reproduce through offsets, the young seedlings can be utilized for vegetative multiplication. Banik (1985) stressed the need for detailed scientific studies on macro-proliferation of bamboo seedlings to develop this into a new dependable technique for bamboo propagation. He further suggested that seedling multiplication should not be continued for a long time (e.g. not more than 10 years in *Bambusa tulda*) because of the problems caused by flowering as the time gap between the last multiplication and flowering gets shorter. The propagation of *Dendrocalamus strictus* through macro-proliferation was studied by Kumar *et al.* (1988a). It was found that the mother stock of seedlings can be multiplied four to six times, depending on fertilizer application in a period of 8 months. The proliferated seedlings remain small in size, due to continuous rhizome separation, and are easy to handle and transport.

Rhizomes

Two general types of rhizomes are found in bamboos: first the pachymorph rhizome (sympodial) is thick, fleshy and short and determinate in growth.

The plant appears as a many-branched clump, growth continuing only from the lateral branches of the rhizome. The rhizome grows horizontally with roots arising on the lower side. The second type, the leptomorph rhizome (monopodial), is slender with long internodes. It is indeterminate and grows continuously in length from the terminal apex and the lateral branch rhizomes. The culm is symmetrical and has lateral buds at most nodes, nearly all of which remain dormant. This type does not produce a clump but spreads extensively over an area. The rhizomes are cut or broken into pieces, and adventitious roots and new shoots develop from the nodes. When planted, rhizomes, with at least one developed bud, produce mature culms in about half the time taken by seedlings.

In species with leptomorph rhizomes, which are found in temperate regions only, Oh and Aoh (1965) found that planting 40–50 cm long rhizome cuttings 10 cm deep gave good results. This method has become the normal practice in Korea where 1000 rhizome cuttings are planted per hectare. In species with pachymorph rhizomes (commonly found in tropical and subtropical regions), offsets are used but are bulky, heavy and difficult to handle and transport. Because only a limited supply of the planting material needed for offsets is available per clump this method is impractical for use in large plantation programmes.

Offsets

Segments (approximately 1 m long) of 1–3 year old culms are used for planting. These consist usually of a portion of the old culm with rhizome and roots, and are cut off at a node 30–60 cm above the ground. The slender tips of the culms are discarded. The offsets are planted horizontally at least 2.5 cm deep in trenches. The single node cutting is planted slightly slanting, the node being underground. Planting is done in the third week of June, and shade is provided against the direct sun to avoid rapid evaporation (Dabral, 1950).

Peal (1882) showed that propagation of bamboos by offset planting was common in the villages of Assam and Bengal. Hasan (1977) reported that the success rate for establishment of offsets was 5% in *Melocanna baccifera*, 9% in *Bambusa tulda*, 33% in *Oxytenanthera nigrociliata* and 40% in *Dendrocalamus longispathus*. Generally, part-clump planting of *Melocanna baccifera* shows better success (35%), as this type of planting material has more than one rhizome with many buds, than offsets which have only one rhizome and a limited number of buds (Banik, 1984b). The rate is better (reaching 100%) in some thick-walled bamboo species (e.g. *Bambusa vulgaris*).

Layering

This is the more usual method of propagation with *B. tulda* and *B. vulgaris*; the layers are made by partly cutting a culm and layering in the soil for rooting. When the shoots appear, the internodes are cut, and the layer is planted separately. The cuttings can be raised in several ways i.e. in nursery, in stagnant water, floating in a pond, and in the bed of water-courses.

Marcotting (air-layering)

This involves bending a 1-year-old culm and making an undercut at its base. The branches at the nodes are pruned to about 2.5 cm in such a way that the dormant buds are not injured. A mixture of garden soil and leaf mould, placed around each node, is wrapped longitudinally with coconut fibre. This is then securely tied at both ends.

Cabanday (1957) tried to propagate *Bambusa blumeana* by using four different methods: (1) by unsplit cutting; (2) by split cutting; (3) by ground layering; and (4) by marcottage. He obtained a maximum survival of 69.9% by air layering or marcottage.

Culm and branch cuttings

With a few exceptions, every node of every segmented axis of a bamboo plant bears a bud or a branch, and a branch, in turn, has a bud at nearly every node. Several studies on vegetative propagation have aimed at transforming as many as possible of these inumerable buds into planting materials (Banik, 1980). Cuttings usually consist of one or more inter-nodes, the lowest bearing root-buds capable of growing out.

Culm cuttings

Bamboos vary greatly in their ability to root from culm cuttings. Troup (1921) reported on the relationship between rooting ability and the abundance of roots on the culm. In 1966 McClure published a method relating rooting ability to root abundance on central branch bases in the mid-culm region.

Investigations on vegetative propagation from cuttings were initiated as early as the nineteenth century when Pathak (1899) tried propagation of *Dendrocalamus strictus* using 3–5 year old cuttings. Curran and Foxworthy (unpublished 1912, cited in Bumarlung and Tamorlang, 1980) reported that cuttings could be used for propagation of the Philippine bamboos *Bambusa blumeana, B. vulgaris* and *Gigantochloa birs* with 34, 32 and 6% success, respectively.

According to Dabral (1950), a culm of *Bambusa tulda* was cut about 23 cm (9 inches) from ground level and driven as a stake 0.5 m (1½ feet) deep into the ground in June. By the end of the rains in October, a few new green leaves had developed from the node just above the ground. On digging up the stake, development of roots from the buried nodes was observed. This was investigated further with *Bambusa arundinacea, B. polymorpha, B. tulda, Dendrocalamus longispathus, D. strictus* and *Thyrostachys oliveri.* One metre (3 feet) segments of culms consisting of at least three nodes were obtained from the clumps of each species. The branches were trimmed to 7–10 cm (3–4 inches) in length. In some cases single nodes were also cut. Culms older than 3 years were avoided. The segments were planted horizontally in the soil. By July sprouting was noticed in most of the bamboos from the nodes, and development continued until the end of October. The most promising results were obtained with *Dendrocalamus longispathus, Thyrostachys oliveri, Bambusa tulda* and *B. polymorpha. B. arundinacea* and *Dendrocalamus strictus,* the most important Indian bamboos, gave a poor response. Lin (1962, 1964) and Chinte (1965) tried using culm segments in a similar way. However, their observations were limited to the short period when they observed the development of sprouts from the nodes: they did not ascertain the development of the rhizome, classifying bamboos as easy to propagate by culm segments.

Culm segments 0.5–1.0 m long were used by Banik (1984a) for propagating bamboos. The cuttings were generally placed slanting at about 45°, 7–15 cm deep in any rooting medium (preferably coarse sand). An experiment showed that 45–56% of the cuttings of different thick-walled bamboos such as *B. vulgaris, B. polymorpha* and *D. giganteus* and 38% of *B. nutans* gave successful propagules. Culm cuttings of thin-walled bamboo species like *Melocanna baccifera, B. tulda, D. longispathus* and *Oxytenanthera nigrociliata* failed to produce any propagules. (Some success was achieved in both air-layering and ground-layering. About 10% of the branches/nodes in a culm produced rooted propagules in *B. vulgaris* and *B. giganteus* (Banik, 1984b; Serajuddoula, 1985) whereas *M. baccifera* did not respond to any of the layering methods.)

In an earlier study, entire 2-year-old culms of *B. nutans* with small branches were severed above the rhizome in March and planted in loamy beds. The environment was improved by shading, irrigation and by using sealed polythene tunnels. Evaluation after several months showed an overall mean production of only 2.2 roots/plant, with successful plants arising from only 8.8% of the nodes. No significant differences were recorded between the treatments (McClure and Kennard, 1955). They obtained a higher percentage of plants from four other *Bambusa* species which suggested that *B. nutans* is very reluctant to root. This view is strongly supported by its morphological features. It is of fine form and has no trace of aerial root production or of the nodal swelling associated with it. In contrast to the above results, *D. hamiltonii* produced rooted shoots from 70% of the nodes while *D. hookeri* produced them from 84% of the nodes. Both of these *Dendrocalamus* species produced the most abundant rooting from the central branch if it had a bud at planting, or from shoots from the basal bud if this had already been developed. Shoots oriented horizontally produced more roots than those which were vertical.

Bearing in mind the limitations of adequate rooting in *B. nutans*, a new planting technique was developed for *Bambusa* species. Two-year-old culms with a reasonably strong central branch were selected. The central branches were only cut beyond the first elongated internode, 15–25 cm from the culm, while other branches were cut at 2–4 cm both to promote development of shoots from the central branch and to simplify orientation of the cutting. The overall success rate was 59.5% with an average of 11.3% plants per culm. Excluding the cuttings which came from the upper regions of the culm where its diameter was less than 3.5 cm, the success rate was 64.5% with a productivity of 10.7% rooting nodes per culm. This showed that by reducing the effects of the factors seen to limit rooting in *B. nutans* and planting single-node cuttings in the correct orientation, a satisfactory response could be obtained in this very similar species which has no aerial roots at all. The stimulation of production of only a few roots on certain shoots, allowed the development of strongly rooted shoots from the basal buds, giving plants in a predictable and fairly uniform fashion. Further studies on the removal of small shoots suggested that these had not been diverting resources excessively. Cuttings with two shoots developing downwards had a greater potential for producing roots than those with only one (Stapleton, 1987).

Abeels (1961) tried different methods of layering of stems, raising cuttings of *Bambusa vulgaris* var. *striata*, *B. viridis-glaucescens*, *Dendrocalamus strictus*, *Gigantochloa apus*, *G. aspera* and *Oxytenanthera abyssinica* in a nursery, in stagnant water, in ponds, in the bed of a watercourse. It was found that only cuttings of *B. vulgaris* var. *striata* produced shoots and roots after the fourth week in a nursery, while cuttings of *B. viridis-glaucescens* rooted after 5–6 weeks in the pond. The same species rooted in the bed of a watercourse after 3 months. The other species did not respond.

It is clear that bamboo cuttings must develop shoots, roots and rhizomes for success. Hasan (1977) tried vegetative propagation through offsets and branch cuttings on a few bamboo species of Bangladesh and estimated the percentage of failure in each phase (rooting and rhizome development). When branch cuttings were used, many species showed good rooting response (0.3–100%) but poor rhizome development. Ultimately in his study, only very few (0.3–1.1%) of the species producing roots also produced rhizomes. The failure rate was somewhat less for the offset method, as the planting material already possessed rhizome and root initials.

To make the best use of the limited cuttings, these must be planted in a way which results in the greatest number of plants. In a comparison of different orientations of planting, horizontal cuttings proved to be more successful than vertical or oblique cuttings (Medina *et al.*, 1962). Gupta and Pattanath (1976) also observed that it was only in the horizontal method that rhizome formation was initiated in the case of *D. strictus*. A comparison of whole culms, two node cuttings and single node cuttings showed that single node cuttings gave the maximum number of shoots per culm (Cabanday, 1957; Dai, 1981).

Effect of seasonal variation

The importance of the physiological state of the culm for subsequent shoot production has also been pointed out (Gupta and Pattanath, 1976). They found that in *D. strictus* there was significant seasonal variation in the stored nutrients and the performance of the propagules varied significantly with different months of planting. April was found to be the best for the preparation of cuttings. It is known from experience that in other plants, the optimum time is often before growth initiation in the plant's normal cycle. As investigated in *D. strictus* (Gupta and Pattanath, 1976), growth in the branches usually starts several months earlier, early spring being the best for propagation. Dai (1981) is also of the same opinion.

White (1947), studied the effect of seasonal variation on the rooting of branch cuttings of nine different bamboo species, *B. longispiculata, B. polymorpha, B. textilis, B. tulda, B. tuldoides, Cephalostachyum pergracile, D. asper, Gigantochloa apus* and *Sinocalamus oldhamii.* His results showed that the average rooting percentage was at a maximum in December (15.8%), but that it was in March that the maximum number of species was stimulated (six of nine species) including *B. longispiculata, B. textilis, B. tulda, D. asper, S. polymorpha, G. verticillata* and *S. oldhamii.* The percentage of rooting was 10.5 for three species stimulated in June. The poorest response was in September (3.3%) with only three species responding (*B. textilis, D. asper* and *S. oldhamii*).

Effect of age

From the published literature and practical experience, it is evident that there is an optimum age for rooting in each type of propagation material (rhizome, offset and branch). Different species have different optimum ages for taking cuttings (McClure and Kennard, 1955). According to McClure (1966) this could be due to variations in bud development and the physiological materials and food reserves in supporting tissues. For instance one rhizome of *Phyllostachys* species did not produce any new culm even 7 years after planting (Ueda, 1960). As the rhizome becomes older the buds gradually lose their sprouting ability. Practical experience has shown that rhizomes more than 2 years old generally give poor and unsatisfactory results (Banik, 1980). In culm cuttings, age also plays a vital role in propagation. Culm cuttings from *D. strictus* less than 2 years old proved better for propagation (Guha *et al.*, 1976) than older culms.

Effect of environmental conditions

In an experiment with *B. vulgaris*, Khan (1972) observed that the best conditions to plant the cuttings is with the onset of rains, on a cool, rainy or cloudy day in clayey or heavy soils. The bed should be flood irrigated, if it is not a rainy day or the rains are insufficient. Abeels (1961), also pointed out that the method of floating in a pond or planting in sandy beds of watercourses permits an easy and economic preparation of well-rooted cuttings.

Effect of plant growth regulators

White (1947) conducted an experiment on *B. longispiculata, B. poly-morpha, B. textilis, B. tulda, B. tuldoides, Cephalostachyum pergracile, D. asper, Gigantochloa apus* and *Sinocalamus oldhamii*. The basal portions of the cuttings obtained from 2-year culms were dipped in one of the following solutions: (1) 5 mgml^{-1} of indole acetic acid (IAA); (2) 2 mgml^{-1} of indole butyric acid (IBA); (3) 2 mgml^{-1} of napthalene acetic acid (NAA) or (4) 1 mgml^{-1} of 2,4-dichlorophenoxyacetic acid (2,4-D). Observations were made 2 months after planting. No differences in rooting were found which could be associated with the root-promoting treatments. Similar results were reported by Delago (1949). However, some reports have emphasized the role of growth regulators in rooting. For example, Uchimura (1977) found that treatment with IBA was better than that with IAA and NAA for rooting of *B. vulgaris*. Suzuki and Ordinaro (1977) reported 45% rooting of IBA-treated two-noded obliquely planted cuttings whereas 80 and 60% success was achieved with *B. vulgaris* and *D. merrilianus*, respectively. Seethalakshmi *et al.* (1983) found that coumarin, NAA and a mixture of coumarin and IAA gave the highest percentages of rooting in the case of *B. balcoa*. Surendran *et al.* (1983) found that culm segments treated with coumarin, NAA and boric acid showed better sprouting and rooting than the control in the case of *B. arundinacea*. However, in none of the above reports is any indication given of rhizome induction, which is very important for the ultimate survival of the plant. In one report on *B. tulda* (Kumar *et al.*, 1988b), rhizome formation was achieved by the application of the growth regulators coumarin, NAA and boric acid (10 mgl^{-1} and 200 mgl^{-1} of each compound). Cuttings were treated with growth regulators in summer (May) and during the rainy (August) season. Coumarin at 10 and 100 mgl^{-1} increased rhizome formation to 40% from 20% (in the control) in the May planting, whereas NAA at 10 mgl^{-1} increased rhizome formation to 60%. Boric acid at 10 and 100 mgl^{-1} increased rhizome formation to 40 and 60%, respectively, in the August planting. In another report, Sharma and Kaushal (1985), indicate rhizome formation. However, in this case, basal cuttings were used which already contained preformed rhizome primordia.

Seethalakshmi *et al.* (1983) carried out experiments on vegetative propagation of culm cuttings of the bamboo reeds, *Ochlandra travancorica* and *O. scriptoria* at monthly intervals from June 1981 to May 1982. For enhancing the rooting response, four growth regulating substances viz. IAA, IBA, NAA and coumarin were tested. Observations on rooting number and height of shoots were recorded after 6 months. In both species, rooting occurred only during February to June. The most promising treatments for *O. travancorica* were coumarin (10 p.p.m.) and NAA (100 p.p.m.) in April (50% rooting), whereas IBA (100 p.p.m.) in March was effective for *O. scriptoria* (50% rooting).

Branch cuttings

Rivière and Rivière (1879) pointed out the similarity between the swollen central branch base and the rhizome, and its potential value in vegetative propagation in many genera. Abeels (1961) who tried branch cuttings, also came to the conclusion that as long as the swollen basal portion of the branch was present, the planting medium did not matter. Cobin (1947) reported the successful use of branch cuttings to propagate *Sinocalamus oldhamii, B. vulgaris* var. *vittata* and *G. verticillata.* In these three species, root primordia appear simultaneously in abundance on the swollen part of the principal branch base. McClure and Durand (1951) point out that, in the matter of striking roots, the basal buds are slow to sprout and during this time the planted materials often die, with the result that the percentage of success is low. Previous studies showed that the ultimate establishment (rhizome development) of normal branch cuttings in bamboo was poor even after abundant root production (60–75%) by rooting hormone (White, 1947; Abeels, 1961; Hasan, 1977). In most cases bamboo cuttings rooted well with hormonal application but the majority of them did not produce any new culm mainly due to the failure of rhizome development (Hasan, 1977; Banik, 1980). Thus although using branch cuttings instead of offsets overcame the difficulties of scarcity, bulk and weight of planting material, the success of propagation was very limited.

Several researchers (Chaturvedi, 1947; McClure, 1966; Banik, 1980) have stressed the importance of selecting branch cuttings that have spontaneous *in situ* rooting and rhizome tips at their base. Artificial induction of the rhizome is possible by chopping the culm tops and removal of newly emerging culms (Banik, 1984a). Regular removal of emerging culms produced more (3.4–83.8%) pre-rooted and pre-rhizomed branches per bamboo clump than chopping the culm top (9.1–27.3%). Such cuttings perform better than normal branch cuttings (Banik, 1984a). Normal branch cuttings require 6–12 months for rooting and 12–30 months for rhizome development (Hasan, 1977). The thin-walled bamboo species also did not show any promising results with branch cuttings or with layering and cutting techniques (Hasan, 1980).

In vitro propagation

Plant tissue culture offers unique opportunities not only for realizing the totipotency of cells into whole plants but also provides conditions under which physiological manipulations can be carried out with the objective of overcoming endogenous controls inherent in the intact plant. *In vitro* methods offer an attractive alternative to offsets, cuttings and seeds for the propagation of bamboos. Although development of tissue culture technology, its field-testing and refinement of procedures may be a slow and time-consuming process, when once established, it enables mass-production of plantlets on an industrial scale.

Several thousands of bamboo plantlets produced using somatic embryogenesis tissue culture techniques are now growing in the forests. Somatic embyrogenesis is defined as embryo initiation and development from cells that are not products of gametic fusion. Thus hundreds of plantlets can be formed from embryoids. Another method, the technique of micropropagation (or *in vitro* vegetative propagation), can yield millions of replicates of an original parent plant. In bamboos, the minor nodes bear axillary buds which remain dormant most of the year and sprout generally during the rainy season; these buds have the capacity to transform into complete plantlets (McClure, 1966). This has been achieved in tissue culture, though with uneven success in some bamboos (Tikiya, 1984; Roohi, 1989). Jerath (1986) has indicated the possibility of inducing rhizomes in these buds in tissue culture.

Several other tissue culture methods, such as precocious rhizome induction are also possible. Treatment of seeds with various hormones, in addition to increasing germination, induces rhizome formation in a majority of plants. Other methods such as detopping used *in vitro*, also lead to precocious rhizome induction. This is very useful as it facilitates the easy transfer of plantlets to soil.

Table 4.2 lists the bamboo species that have been used for plant tissue culture. The availability of explants all the year round for tissue culture is not a problem. Moreover diverse explants can be utilized. Table 4.3 lists the explants that can be used for tissue culture work. The multiplication rate in tissue culture is also high and is often several times that obtained in conventional methods. Besides, only a limited number of explants/clumps

Table 4.2. Bamboo species used for plant tissue culture.

Bambusa arundinacea
B. beecheyana var. *beecheyana*
B. glaucescens
B. multiplex
B. oldhamii
* *B. ventricosa*
* *B. vulgaris*
Dendrocalamus latiflorus
D. strictus
Phyllostachys aurea
P. viridis
Sasa pygmae
Sinocalamus latiflora
* *Schizostachyum brachycladum*
* *Thyrostachys siamensis*

---

* Regeneration of plantlets not yet achieved

can be obtained for conventional propagation. Thus, propagation by conventional vegetative methods does not make efficient use of the resource. The explants are also available after long periods in the case of conventional methods, whereas in the case of *in vitro* methods, explants are available as and when required, all the year round. Most important, tissue culture methods are entirely independent of the external environment and can be

Table 4.3. Explants used for tissue culture.

Explant	
Inflorescence	
Immature embryo	
Mature embryo	Plantlet through somatic embryogenesis
Seedling leaf sheath	
Seedling root	
Seedling rhizome	
Seedling nodes	Plantlets through somatic embryogenesis and micropropagation
Seedling basal node	
Mature node	Shoot formation. Rooting of shoots achieved only in a
Shoot	small percentage of cultures. Callusing also achieved
Mature rhizome	but without embryogenesis

Table 4.4. Handling of conventional and tissue culture raised bamboo propagules.

Source of plants	Shipping weight/plant	Planting care
Rhizome offsets	20–40 kg	Suitable for field planting
Culm cuttings	3–8 kg	Suitable for field planting
Tissue culture raised plantlets		
1 month	7.5–10 g	Need nursery care
12 months	100 g	Suitable for field planting

carried out all the year round, whereas conventional vegetative methods are limited by the growth cycle of the plant as well as environmental conditions. Handling of the explants also poses a problem due to the heavy weight of the explant, thus making the process very tedious and difficult. The weight of the rhizome offsets being as much as 20–40 kg, these are difficult to handle and expensive to transport. In comparison, the weight of the tissue culture derived plants is much lower (Table 4.4).

Thus to ensure adequate planting material, the use of tissue culture techniques and the coupling of plant physiological approaches to conventional propagational methods [e.g. treatment of cuttings with several growth regulators etc. (Surendran and Seethalakshmi, 1985)] is advocated for the interim. With more improvements in tissue culture techniques, e.g. development of the technology of suspension cultures and secondary somatic embryogenesis, the future belongs to tissue culture, especially with an in-built industrial approach. Embryogenic suspension cultures offer the prospect of large scale cloning of plants through proper staging and control of development, the imposition of artificial dormancy and the creation of artificial seeds and/or the use of mechanized delivery systems.

Acknowledgements

We thank Mr Faiyaz Q. Shamsi and Mr K. Gangadharan Pillai for their help in the preparation of this manuscript. The tissue culture work on bamboos in our laboratory was funded by the Department of Biotechnology, Government of India.

References

Abeels, P. (1961) Propagation of bamboo. *Bulletin of Agricultural Congress* 52, 591–8.

Banik, R.L. (1980) Propagation of bamboo by clonal methods and by seed. In: Lessard, G. & Chouinard, A. (eds), *Bamboo Research in Asia.* Proceedings Workshop, Singapore. IDRC, Canada, pp. 139–50.

Banik, R.L. (1984a) Macropropagation of bamboos by pre-rooted and pre-rhizomed branch cutting. *Bano Biggyan Patrika,* Forest Research Institute, Chittagong, Bangladesh, 13, 67–73.

Banik, R.L. (1984b) Studies on the propagation techniques of different bamboo species of Bangladesh. Paper presented at *The Workshop in the Contract Research Project,* Bangladesh Agriculture Research Council (BARC). 17–20 November, Dhaka, Bangladesh.

Banik, R.L. (1985) Management of wild bamboo seedlings for natural regeneration and afforestation programme. *Proceedings of 10th Annual Bangladesh Science Conference.* Dhakar, Bangladesh, pp. 78–9.

Bumarlang, A.A. & Tamorlang, F.N. (1980) Country report of Philippines. In: Lessard, G. & Chouinard, A. (eds), *Bamboo Research in Asia.* Proceedings Workshop, IRDC, Canada, pp. 69–80.

Cabanday, A.C. (1957) Propagation of Kanayan-tinik (*Bambusa blumeana* schultz.) by various methods of cutting and layerage. *Philippine Journal of Forestry* 13, 81–97.

Chaturvedi, B. (1947) Aerial rhizome in bamboo culms. *Indian Forester* 73, 543.

Chinte, F.O. (1965) Bamboos in plantation. *Forestry Leaves* 16, 33–9.

Cobin, M. (1947) Notes on the propagation of sympodial or clump type bamboos. *Proceedings Florida State Horticultural Society* 60, 181–4.

Dabral, S.N. (1950) Preliminary note on propagation of bamboos from culm segments. *Indian Forester* 76, 313–14.

Dai, O.H. (1981) Raising plants of bushy bamboos from branched culms with notched internodes. *Forest Science and Technology (Linye Keji Tongxun)* No. 1, 3–6 (Chinese).

Delago, R.F. (1949) Rooting side branch cuttings. *Report of Federal Experimental Station, Puerto Rico,* 24.

Guha, S.R.D., Singh, M.M. & Bhola, P.P. (1976) Beating characteristics of bamboo pulp in velley beater: effect of temperature and consistency on power consumption and pulp sheet properties. *Ippta* 13, 49.

Gupta, B.N. & Pattanath, P.G. (1976) Variation in stored nutrients in culms of *Dendrocalamus strictus* and their effect on rooting of culm cuttings as influenced by their method of planting. *Indian Forester* 102, 235–41.

Hasan, S.M. (1977) Studies on the vegetative propagation of bamboos. *Bano Biggyan Patrika,* Forest Research Institute, Chittagong, Bangladesh 6, 64–71.

Hasan, S.M. (1980) Lessons from the past studies on the propagation of bamboos. In: Lessard, G. & Chounard, A. (eds), *Bamboo Research in Asia.* Proceedings Workshop, Singapore. IDRC, Canada, pp. 131–8.

Jerath, R. (1986) '*In vitro* propagation of bamboo.' Unpublished MPhil thesis, University of Delhi, Delhi, India.

Khan, M.A.W. (1972) Propagation of *Bamboo vulgaris* – its scope in forestry. *Indian Forester* 98, 359–62.

Kumar, A., Gupta, B.B. & Negi, D.S. (1988a) Vegetative propagation of *Dendrocalamus strictus* through macropropagation. *Indian Forester* 114, 564–8.

Kumar, A., Dhawan, M. & Gupta, B.B. (1988b) Vegetative propagation of *Bam-*

busa tulda using growth promoting substances. *Indian Forester* 114, 569–75.

Lin, W.C. (1962) Studies on the propagation by level cutting of various bamboos. *Bulletin Taiwan Forestry Research Institute* 80, 48 (Chinese with English titles and summary).

Lin, W.C. (1964) Studies on the propagation by level (horizontal planting) cutting of various bamboos. *Bulletin Taiwan Forestry Research Institute* 103, 58 (in Chinese).

McClure, F.A. (1966) *The Bamboos: A Fresh Perspective.* Harvard University Press, Cambridge, Massachusetts.

McClure, F.A. & Durand, F.M. (1951) Propagation studies (with bamboo). *Report of Federal Experimental Station, Puerto Rico.*

McClure, F.A. & Kennard, W.C. (1955) Propagation of bamboo by whole-culm cuttings. *Proceedings American Society Horticultural Science* 65, 283–8.

Medina, J.C., Ciaramello, D. & Catro, G.A. (1962) Propagacao-vegetativa do bambu imperial (*Bambusa vulgaris* Schrad var. *vittata* A. et C. Riv.) *Bragantia (Boletin Technico do Institute Agrenomico de Sao Paulo)* 21 (37), 653–65 (in Portuguese).

Oh, S.W. & Aoh, Y.K. (1965) Influence of rhizome length and transplanting depth upon survival of bamboo root cuttings. *Research Report,* Office of Rural Development, Seoul, S. Korea 8, 41–42.

Pathak, S.L. (1899) Propagation of the common male bamboo by cuttings in the Pinjaur Patiala forest nurseries. *Indian Forester* 25, 307–8.

Peal, S.E. (1882) Bamboo for paper stock. *Indian Forester* 8, 50–4.

Rivière, A. & Rivière, C. (1879) Les bambous. *Bulletin Societe National d'Acclimation de France* 5, 221–53, 290–322, 392–421, 460–78, 501–26, 597–645, 666–721, 758–828.

Roohi, F.N. (1989) 'Propagation of bamboos.' Unpublished MPhil thesis. University of Delhi, Delhi, India.

Seethalakshmi, K.K., Venkatesh, C.S. & Surendran, T. (1983) Vegetative propagation of bamboos using growth promoting substances. I. *Bambusa balcoa* Roxb. *Indian Journal of Forestry* 6, 98–103.

Serajuddoula, M. (1985) Propagation of *Bambusa vulgaris* Schrad and *Melocanna baccifera* Trin. by layering. *Proceedings 10th Annual Bangladesh Science Conference.* Dhaka, Bangladesh, pp. 79–80.

Sharma, O.P. & Kaushal, S.K. (1985) Exploratory propagation of *Dendrocalamus hamiltonii* Munro by one-node culm cuttings. *Indian Forester* 111, 135–9.

Stapleton, C.M.A. (1987) Studies on vegetative propagation of *Bambusa* and *Dendrocalamus* species by culm cuttings. In: Rao, A.N., Dhanarajan, G. & Sastry, C.D. (eds), *Recent Research on Bamboos. Proceedings International Bamboo Workshop* 6–14 Oct., 1985), Hangzhou, Peoples Republic of China, pp. 146–53.

Surendran, T., Venkatesh, C.S. & Seethalakshmi, K.K. (1983) Vegetative propagation of the thorny bamboo *Bambusa arundinacea* (Retz.) Willd. using some growth regulators. *Journal of Tree Science* 2, 10–15.

Surendran, T. & Seethalakshmi, K.K. (1985) Investigations on the possibility of vegetative propagation of bamboos and reeds by rooting stem cuttings. *KRFI Research Report,* 31.

Suzuki, T. & Ordinaro, F.F. (1977) *Some Aspects and Problems of Bamboo*

Forestry and Utilization in the Philippines. Asia Forest Industries, College, Laguna, Philippines.

Tikiya, N. (1984) '*In vitro* propagation of the golden bamboo'. Unpublished MPhil Thesis. University of Delhi, Delhi, India.

Troup, R.S. (1921) *The Silviculture of Indian Trees.* Vol II. Clarendon Press, Oxford.

Uchimura, E. (1977) *Ecological Studies on the Cultivation of Bamboo Forest in Philippines.* 74. Forest Research Institute Library, College, Laguna, Philippines.

Ueda, K.Z. (1960) *Studies on the Physiology of Bamboo with Reference to Practical Application.* Prime Minister's Office, Resources Bureau, Science & Technics Agency, Tokyo, Japan, Reference data, 37, 167.

White, D.G. (1947) Propagation of bamboo by branch cuttings. *Proceedings of the American Society for Horticultural Science* 50, 392–4.

Chapter 5

Rapid Propagation of Bamboos through Tissue Culture

I. Usha Rao, I.V. Ramanuja Rao and Vibha Narang

*Department of Botany, University of Delhi, Delhi
110007, India*

Bamboos are one of the most unique and versatile groups of plants known to humankind and perhaps for this reason, have been intimately associated with the civilization of man since ancient times. It is said that one-third of humanity uses bamboo in one form or another during their lifetime. In the modern industrial society which is characterized by plastics and steel, bamboos are still relevant and contribute to human well-being.

Availability of bamboos

At one time the supply of bamboo was thought to be perpetual. The social use of bamboos made little demand on the resource and was far short of the annual increment. Thus, bamboo was often viewed by foresters in many countries as a weed species and a nuisance due to its rapid growth and, therefore, its mass utilization in the pulp and paper industries was welcome.

In India, the paper industry is largely dependent upon bamboo which is preferred over other softwoods and hardwoods as a raw material. It is estimated that 6–7 times as much cellulose material can be obtained from a hectare of bamboo forest as can be had from other broadleaved or coniferous forest (Varmah and Bahadur, 1980). The principal sources of paper pulp of acceptable quality in India are *Dendrocalamus strictus* and *Bambusa arundinacea.* It is estimated that out of an annual production of nearly 9.5 million metric tonnes of bamboo in India, about 4.9 million tonnes are presently utilized for paper making (Varmah and Pant, 1981). This yields around 600000 tonnes of paper pulp per year whereas the country's requirement in 1984 was 3.5 million tonnes, which may rise to 4.5 million tonnes by the turn of the century (Varmah and Pant, 1981).

In India, industrial production of paper started in the 1930s. Since independence, however, there has been a rapid expansion of paper mills with 35 of them using bamboo as a source of long fibre (Liese, 1985). After the Second World War, Sri Lanka and Indonesia also set up paper and pulp factories. China recently set up several small-scale paper mills using bamboo pulp (Jifan, 1987).

Whereas this modern usage of an ancient plant was a significant advance, it also caused the decimation of the vast stocks of bamboo in the country. The resulting severe shortage of raw material for paper production and its other traditional uses has necessitated a replanting programme. This brought into focus the hitherto unexperienced problem of availability of planting material.

Conventional propagation of bamboos

Seeds

Bamboo propagation occurs by and large through offsets and seeds. Seeds are very convenient for plant propagation and due to their small size, are easily transportable. Bamboo seeds are light in weight: 1 kg of *Bambusa arundinacea* seeds contains about 90000 seeds; *Bambusa tulda*, 26000; *Dendrocalamus longispathus*, 150000; *Dendrocalamus strictus*, 40000 and the pear-like seeds of *Melocanna bambusoides*, only 70 (Liese, 1985).

Prasad (1986) reported that the average seed production in gregariously flowered areas in the Shahdol district of Madhya Pradesh State (India) varied from 0.9 tonne to 1.5 tonnes ha^{-1}. Although gregarious flowering results in the production of a large number of seeds, propagation by seeds is not always practical because of the unusually prolonged flowering cycle of monocarpic bamboos. This is commonly the major limitation. For example, *D. strictus* is known to flower gregariously at intervals of around 40–50 years. Sporadic flowering, or off-cycle flowering, which occurs almost every year, does not usually result in seed formation, or at best in only a few viable seeds from a large mass of empty florets. However, due to the relatively large number of flowering cohorts, some seeds are always available, although the quantities may vary greatly from year to year. Such a situation is largely the case for *D. strictus*, and less so for *B. arundinacea*. Thus, seeds are generally not available or are in short supply for most bamboos. This is indirectly supported by the fact that over the years, vegetative propagation methods have been developed in various communities in several countries. In comparison, seed-based propagation is either unknown or nearly non-existent.

The difficulty in using bamboo seeds lies in their low viability, poor storage characteristics and inborne microbial infection. In addition, the

dependence on sporadic off-cycle flowering makes their availability irregular. The principal problem with the use of bamboo seeds is their poor
viability (Bahadur, 1979; Varmah and Bahadur, 1980). Generally, bamboo
seeds loose viability about 2–3 months after harvest. Another problem is
the presence of seed-borne fungi and other microbes in some bamboo
species. The embryos in several such infected seeds do not germinate.

Because of the abundant seed production in gregarious flowering,
sufficient regeneration usually occurs except where the soil is too hard. In
most places, seedlings appear in large numbers. However, grazing and fire
are most detrimental to the survival of seedlings. In order to overcome this
problem, the entire flowering area should be closed but this is not always
possible due to socio-political problems. In an experiment in which half of
the flowering area in Shahdol region (Madhya Pradesh, India) was closed
for grazing in each felling series (Dwivedi, 1990), it was observed that in
unprotected areas regeneration was poor and the seedlings generally
existed only in places where some physical and mechanical barriers were
provided. In protected areas enough seedlings existed to regenerate the
area.

Vegetative propagation

Various methods of vegetative propagation such as offset planting, rooting
of culm and branch cuttings, and layering are used for propagation of
bamboos. The success and limitations of such methods have been reviewed
by Banik (1980) and Hasan (1980). The advantage of vegetative propagation is that the genetic quality of the planting material is known because it
is the same as that of the source plant. A common drawback of all vegetative propagation methods is that these can be carried out successfully only
in certain seasons.

Offsets

Vegetative propagation by offsets has proved to be of limited value. A
common method is to cut off part of the rhizome from a young shoot (1–2
years) and transplant this portion of the clump into a prepared hole. A 30–
50 cm portion of the rhizome with a number of nodes and buds may be
used if the clumps are dense with rhizomes and the culms closely inter-
laced. With adequate wetting, a new shoot appears. This method has been
commonly employed for species with leptomorph rhizomes (e.g., *Phyllostachys* and *Melocanna*). These expand by a runner-type extension giving
rise to groves or large tracts. In the evenly warm tropical region, clump
type stands are formed. The rhizomes are very short, solid and often
thicker than the culm. The culms are densely clustered and propagation is
difficult as offsets are bulky, heavy and difficult to handle and transport.

Digging up the rhizomes entails considerable risk of damage to the parent plant. Also only a limited supply of the offset planting material is available per clump and, therefore, this method is impractical for use in large plantation programmes.

Culm cuttings

Bamboos vary greatly in their ability to root from culm cuttings. Generally, culm cuttings 0.5–1.0 m long are used for propagating bamboos (Banik, 1984). Culm cuttings are commonly placed in a slanting position, 7–15 cm deep in any rooting medium (preferably coarse sand). It has been observed (Banik, 1984) that 45–56% cuttings of different thick-walled bamboos such as *B. vulgaris, B. polymorpha* and *D. giganteus* and 38% of *B. nutans* gave successful propagules. Culm cuttings of thin-walled bamboo species such as *Melocanna baccifera, B. tulda, D. longispathus* and *Ochlandra nigrociliata* failed to produce any propagules. Seethalakshmi *et al.* (1990) developed suitable methods for the propagation of bamboo reeds by rooting culm cuttings after treatment with growth regulating substances.

Branch cuttings

Propagation of bamboo through branch cuttings could be a useful approach because of their availability and ease in handling. Previous studies have, however, shown that the ultimate establishment (rhizome development) of normal branch cuttings in bamboo is poor even after abundant root production (60–75% by application of a rooting hormone: White, 1947; Abeels, 1961; Hasan, 1977). In most cases, the bamboo cuttings rooted well with hormonal application but the majority of them did not produce any new culm mainly due to failure of rhizome development (Hasan, 1977; Banik, 1980). However, in *D. asper* and *B. vulgaris,* propagation using branch cuttings is successful and widely practised.

Other methods are marcotting and layering. These are not commonly practised.

In general, it has been observed that there is increasing difficulty in producing bamboo propagules as one goes from the rhizome to the culm and to the branch. At the same time, the number of 'cuttings' or potential propagules increases. McClure clearly foresaw this and called for the development of new methodologies which would utilize this enormous potential for the benefit of mankind. In 1966, he wrote: 'No published account of the successful use of artificial means to break the dormancy buds at will has come to my attention. A satisfactory degree of success in the vegetative propagation of bamboo can be achieved only when routines effectively solving this problem have been established.' To this day, propagation of bamboo from the nodes of the minor branches taken from adult plants remains a difficult if not impossible task. In comparison, there has been marked success using seedling nodes in tissue culture.

While vegetative propagation ensures clonal fidelity, it is necessary to be aware of the debilitating effects of continued vegetative propagation on bamboo production. In Thailand, for example, mass-flowering of the vegetatively propagated *D. asper* is threatening a lucrative export industry. Clearly, the solution to the problem lies somewhere in between and is also related to the scale of the demand. The industrial appetite for bamboo raw material which has not been met using conventional propagation methods and which has come into conflict with the more pressing social needs of the people, needs to be addressed using newer methods that lend themselves to the industrial scale. At the same time, the time-tested conventional techniques will have to be used in smaller scale operations.

In vitro propagation

McClure (1966) wrote:

'Reduction in the mass of the individual propagule makes for economy of the propagating material, simplifies the labour of preparing it, and reduces the requirements of space and other facilities. As the bud is deprived more and more completely of the maternal tissue that supports it, the control is perfected, the number of unassessed and uncontrolled variables is reduced, and the prospects of establishing pertinent basic principles and determining the optimal conditions of vegetative propagation for each kind of bamboo improve.'

Prophetically, McClure (1966) realized that

'the development, or adaptation, or the appropriate refinements of these procedures will require experience in the routines of sterile culture, tissue culture, the breaking of dormancy in buds, the use of hormones for stimulating root initiation, and so forth.'

It has now become possible routinely to produce several thousands of bamboo plants through tissue culture principally through the method of somatic embryogenesis and also by micropropagation.

Somatic embryogenesis

One method is the formation of plantlets from somatic embryos. Somatic embryogenesis is defined as the initiation and development of embryoids from cells that are not products of gametic fusion. These somatic embryos are identical to normal zygotic embryos and are equipped with both the shoot and root pole and only need to be germinated to obtain complete plants. Thus, hundreds of plants can be formed from embryoids. To distinguish from the term embryo which normally refers to the zygotic embryo, those produced through tissue culture are referred to as somatic embryos or embryoids. The approach of somatic embryogenesis has proved to be a successful one and plantlets have been regenerated from several bamboos such as *D. strictus* (Rao *et al.*, 1985, 1987, 1989, 1990a,b; Dekkers, 1989; Zamora and Gruezo, 1990); *Phyllostachys viridis* (Hassan and El-Debergh, 1987); *Bambusa beechyana, B. oldhamii* and *Sinocalamus latiflora* (Yeh and Chang, 1986a,b, 1987).

Micropropagation

An alternative method to somatic embryogenesis is the method of micropropagation, which is, in a broad sense, vegetative propagation *in vitro* or under sterile conditions. For example, the growth controls that operate in an intact plant can be broken down or eliminated under *in vitro* conditions, leading to profuse production of several shoots from a single shoot that was the initial explant. These can be separated and rooted to give rise to entire plants. Micropropagation can thus yield millions of faithful duplicates of an original parent clone plant (Rao *et al.*, 1989). In bamboos, this approach has been followed with uneven success by Tikiya (1984), Nadgir *et al.*, (1984), Jerath (1986), Roohi (1989), Saxena (1990) and others.

Production of plantlets through somatic embryogenesis

Explants

The following explants can be utilized for somatic embryogenesis (Rao *et al.*, 1990b).

   1. Juvenile materials:
      (a) zygotic embryo;
      (b) immature embryo;
      (c) seedling parts (node, leaf sheath, root, rhizome);
   2. Tissue-culture raised materials:
      (a) somatic embryo;
      (b) parts of plantlets regenerated from somatic embryos;
   3. Adult (mature) materials:
      (a) node;
      (b) shoot-tip;
      (c) leaf sheath base;
      (d) rhizome.

   After dehusking, mature seeds are washed in 2% Teepol solution

(Shell, India) on a magnetic stirrer for 5 min. The Teepol solution is removed by washing in running tap water for 15–20 min followed by a rinse in distilled water. Sterilization of the material is effected by a 5 min immersion of seeds in chlorine water (saturated chlorine water diluted five times with distilled water) followed by thorough washing in sterile distilled water. Surface moisture is removed with a sterile filter paper and the seeds implanted in tubes containing the inductive medium under aseptic conditions (Rao *et al.*, 1985).

The inductive medium consists of salts and vitamins of B_5 basal medium (Gamborg *et al.*, 1968), 2% sucrose, 0.8% agar (w/v) supplemented with 2,4-dichlorophenoxyacetic acid (2,4-D) at 10 and 30 μmol concentration. The pH of the medium is adjusted to 5.8 before autoclaving at 104 kPa (15 p.s.i.) for 15 min. The cultures are maintained at $27 \pm 2°C$ under continuous illumination (2500 lx) provided by cool-white daylight fluorescent tubes.

Origin and development of callus and somatic embryos

In *D. strictus* it has been observed that somatic embryos originate from compact callus (embryogenic) and not the friable ones (Fig. 5.1). Histological techniques have been utilized to study the origin and development of embryogenic callus and somatic embryos. For this purpose, cultures of seeds in B_5 + 2,4-D (3×10^{-5} mol) were taken after different periods of inoculation. These were dissected and examined in the stereomicroscope; similar cultures were fixed and sectioned to study the histological changes during callusing and embryoid development.

When observed in the initial week of inoculation, callusing was found to be restricted to the embryonal axis and the scutellum remains as such. Initial callus formation takes place from the coleoptile. Its removal reveals the embryonal axis as a white solid central core. Examination of the callusing embryo revealed that the callus masses were all arising from vascular bundles. Before callusing, accumulation of starch was noted in the tissues adjoining the bundle. Callusing was also noted in the tissues adjoining the bundle. Callusing initiates from the cells located towards the base of the bundles which consist of xylem parenchyma and does not take place on the side of the bundle where the fibrous cap is located. The callusing subsequently spreads all around the bundle. The somatic embryos that are formed from the embryogenic callus are often complete with a shoot and root axis, a coleoptilar region and a scutellum.

Secondary somatic embryogenesis has also been commonly observed. The epidermal cells of the developing embryoids divide to give rise to globular masses which later develop into embryoids. If continually maintained on this medium, these embyronal masses proliferate giving rise to a culture which consists entirely of secondary embryos.

Fig. 5.1. Embryogenic compact callus of *Dendrocalamus strictus* showing differentiation of somatic embryos.

Multiplication of embryoids

In comparison with multiplication of embryogenic compact callus followed by differentiation of embryoids, secondary somatic embryogenesis has been found to be more rapid. In this process the primary embryoids proliferate further and give rise to a new crop of secondary embryoids. These arise by the proliferation of the scutellar epidermis. Numerous embryoids develop from the proliferated epidermis. A further crop of embryoids can be obtained from these secondary embryoids and the process continued indefinitely until removed from the nutrient.

Germination of embryoids and rearing of plantlets

The surplus embryoids from a round of embryoid multiplication are transferred for maturation and germination. Here the embryoids are given an equal opportunity to mature and germinate into plantlets (Fig. 5.2). The plantlets formed are removed to a medium that permits further growth or precocious rhizome induction is carried out. The ungerminated em-

Fig. 5.2. Germination of somatic embryos of *Dendrocalamus strictus* into plantlets.

bryoids are then re-transferred for another round of embryoid maturation and germination.

Potting of plantlets and acclimatization

Mature plantlets which are about 8–10 cm tall are ordinarily potted out (Fig. 5.3). The plantlets are removed from the medium and potted directly into a soil:sand:farmyard manure (1:1:1) potting mix in 10 cm diameter pots. These are then maintained in the acclimatization chamber in which temperature, humidity and light conditions are controlled. After an acclimatization period of 2–3 weeks, the plantlets are transferred to the glasshouse or to PVC greenhouses (Fig. 5.4) where they are maintained for another 2 months. Subsequently they are repotted into polyethylene bags and maintained until they are taken by the foresters.

Ordinarily *in vitro* produced bamboo plantlets are handed over to the foresters when they are 8 months old. Planting in the forest is done when the plants are 12 months old. The rhizome system is well-developed by this time and the plants establish easily and show rapid growth (Fig. 5.5).

Fig. 5.3. Mature plantlets of *Dendrocalamus strictus* ready for potting out.

Nodal culture

Micropropagation – nodal explants

McClure (1966) has emphasized that 'with a few exceptions, every node of every segmented axis of a bamboo plant bears a bud or a branch which in turn, has a bud at every node. Theoretically, at least, each one of these buds is a potential plant. Studies in vegetative propagation should include methods for transforming as many as possible of these innumerable buds into little rooted plants.' There is even the possibility of inducing rhizomes in these buds. In such cases, a separate root formation step is not required as the rhizomes are capable of producing both culms and roots, thus giving rise to complete plants. If successfully established, the use of the dormant axillary buds or nodes would make available a large presently unused resource for propagation. Tissue culture technology thus offers the potential ability to raise in a short span of time and in a limited space thousands of plantlets from the nodal regions of the existing clumps, without destroying them (which occurs when offsets are used).

Fig. 5.4. Plantlets of *Dendrocalamus strictus* from tissue culture in a greenhouse.

Fig. 5.5. Plantlets of *Dendrocalamus strictus* from tissue culture growing at a field site.

The results so far (Jerath, 1986; 1984; Roohi, 1989; Tikiya, 1984) have shown that it is relatively easy to get the axillary buds in the nodes of *Bambusa vulgaris* and *Dendrocalamus strictus* to sprout. Multiple shoots can also be obtained. However, it has been difficult to persuade a sufficiently high percentage of them to root. Presently, depending on the species, between 4% and 10% of the nodal shoots can be rooted (Roohi, 1989). This percentage must be increased.

Micropropagation – multiple shoots from seeds

Yet another approach is the formation of multiple shoots from seeds. The multiple shoots can be rooted or subcultured to obtain another set of multiple shoots. It has been observed that multiple shoots are easily induced from zygotic embryos and these can be rooted (Dekkers, 1989; Nadgir *et al.*, 1984; unpublished data). In *Bambusa arundinacea*, 58.7% of the seed cultures formed multiple shoots in a cytokinin supplemented medium (Jerath, 1986).

Conclusion

The use of tissue culture as a method for the mass propagation of bamboos is now established with several thousands of such plants growing in several States of India. The performance of these plants is largely superior to that of seed-raised plants. The precocious production of rhizomes not only enables easy establishment in the field but also ensures protection from grazing and fire which destroy the above-ground parts. Because of the inherent industrial approach and the several advantages that accrue, it is only a matter of time before the propagation of bamboos through tissue culture becomes the method of choice.

Acknowledgement

The tissue culture work on the bamboos in our laboratory was funded by the Department of Biotechnology, Government of India.

References

Abeels, P. (1961) Propagation of bamboo. *Bulletin of Agricultural Congress* 52, 591–8.

Bahadur, K.N. (1979) Taxonomy of bamboos. *Indian Journal of Forestry* 2, 222–41.

Banik, R.L. (1980) Propagation of bamboos by clonal methods and by seed. In: Lessard, G. & Chouinard, A. (eds) *Bamboo Research in Asia.* IDRC, Canada, pp. 139–50.

Banik, R.L. (1984) Macro-propagation of bamboos by pre-rooted and pre-rhizomed branch cutting. *Bano Biggyan Patrika* 13, 67–73.

Dekkers, A.J. (1989) '*In vitro* propagation and germplasm conservation of certain bamboo, ginger and *Costus* species.' Unpublished PhD thesis, National University Singapore, Singapore.

Dwivedi, A.P. (1990) Gregarious flowering of *Dendrocalamus strictus* in Shahdol (Madhya Pradesh) – Some management considerations. In: Rao, I.V.R. Gnanaharan, R. & Sastry, C.B. (eds), *Proceedings International Bamboo Workshop.* Cochin, Kerala, India, pp. 87–97.

Gamborg, O.L., Miller, R.A. & Ojima, K. (1968) Nutrient requirements of suspension cultures of soybean root cells. *Experimental Cell Research* 50, 151–8.

Hasan, S.M. (1977) Studies on the vegetative propagation of bamboos. *Bano Biggyan Patrika* Forest Research Institute, Chittagong, Bangladesh, 6, 64–71.

Hasan, S.M. (1980) Studies on the vegetative propagation of bamboos. In: Lessard, G. & Chouinard, A. (eds), *Bamboo Research in Asia.* IDRC, Canada, pp. 131–8.

Hassan, A.A. & El-Debergh, P. (1987) Embryogenesis and plantlet development in the bamboo *Phyllostachys viridis* (Young McClure). *Plant Cell Tissue & Organ Culture* 10, 73–7.

Jerath, R. (1986) '*In vitro* propagation of bamboos.' Unpublished MPhil Thesis, University of Delhi, Delhi, India.

Jifan, A. (1987) Bamboo development in China. In: Rao, A.N., Dhanarajan, G. & Sastry, C.B. (eds), *Recent Research on Bamboos.* The Chinese Academy of Forestry and International Development Research Centre, Singapore, pp. 24–5.

Liese, W. (1985) *Bamboos – Biology, Silvics, Properties, Utilization.* Deutsche Gesellschaft für Technische Zusammenarbeit (GTZ) GmbH, Eschbnorn, W. Germany.

McClure, F.A. (1966) *The Bamboos: A Fresh Perspective.* Harvard University Press, Cambridge, Mass., USA.

Nadgir, A.L., Phadke, C.H., Gupta, P.K., Parsharami, V.A., Nair, S. & Mascarenhas, A.F. (1984) Rapid multiplication of bamboos by tissue culture. *Silvae Genetica* 33, 219–23.

Prasad, R. (1986) Bamboo plantation. *Forest Bulletin* No. 22 SFRI, Jabalpur, India.

Rao, I.U., Rao, I.V.R. & Narang, V. (1985) Somatic embryogenesis and regeneration of complete plantlets in the bamboo, *Dendrocalamus strictus. Plant Cell Reports* 4, 191–4.

Rao, I.U., Narang, V. & Rao, I.V.R. (1987) Plant regeneration from somatic embryos of bamboo and their transplantation to soil. In: *Proceedings Symposium on Plant Micropropagation in Horticultural Industries: Preparation, Hardening and Acclimatization Processes,* Belgium, pp. 101–7.

Rao, I.V.R., Yusoff, A.M., Rao, A.N. & Sastry, C.B. (1989) Propagation of bamboo and rattan through tissue culture. In: *Proceedings 3rd International*

Workshop on Bamboos. Cochin, Kerala, India.

Rao, I.V.R. & Rao, I.U. (1990a) Tissue culture approaches to the mass-propagation and genetic improvement of bamboos. In: Rao, I.V.R., Gnanaharan, R. & Sastry, C.B. (eds), *Proceedings International Bamboo Workshop.* Cochin, India.

Rao, I.U., Rao, I.V.R., Narang, V., Jerath, R. & Pillai, K.G. (1990b) Mass propagation of bamboo from somatic embryos and their successful transfer to the forest. In: Rao, I.V.R., Gnanaharan, R. & Sastry, C.B. (eds), *Proceedings International Bamboo Workshop* (Nov. 14–18, 1988). Cochin, India, pp. 167–72.

Roohi, F.N. (1989) 'Propagation of bamboos.' Unpublished MPhil thesis, University of Delhi, Delhi, India.

Saxena, S. (1990) *In vitro* propagation of bamboo species by shoot proliferation and somatic embryogenesis. In: *VII International Congress on Plant Tissue and Cell Culture.* Amsterdam, June 24–29.

Seethalakshmi, K.K., Surendran, T. & Somen, C.K. (1990) Vegetative propagation of *Ochlandra travancorica* and *O. scriptoria* by culm cutting. In: Rao, I.V.R., Gnanahara, R. & Sastry, C.B. (eds), *Proceedings International Bamboo Workshop* (Nov. 14–18, 1988). Cochin, Kerala, India, pp. 136–47.

Tikiya, N. (1984) '*In vitro* propagation of the golden bamboo,' Unpublished MPhil thesis, University of Delhi, Delhi, India.

Varmah, J.C. & Bahadur, K.N. (1980) Country report and status of research on bamboos in India. *Indian Forest Records* (NS) 6 (i).

Varmah, J.C. & Pant, M.M. (1981) Production and utilization of bamboos. *Indian Forester* 107, 465–76.

White, D.G. (1947) Propagation of bamboo by branch cuttings. *Proceedings of the American Society for Horticultural Science* 50, 392–4.

Yeh, Meei-Ling & Chang, Wei-Chin (1986a) Somatic embryogenesis and subsequent plant regeneration from inflorescence callus of *Bambusa beechyana* Munor var. *beechyana. Plant Cell Reports* 5, 409–11.

Yeh, Meei-Ling & Chang, Wei-Chin (1986b) Plant regeneration through somatic embryogenesis in callus culture of green bamboo (*Bambusa oldhamii*). *Theoretical and Applied Genetics* 73, 161–3.

Yeh, Meei-Ling, & Chang, Wei-Chin (1987) Plant regeneration via somatic embryogenesis in mature embryo derived callus culture of *Sinocalamus latiflora. Plant Science* 5, 93–6.

Zamora, A.B. & Gruezo, S.S.M. (1990) Micropropagation and potting out of bamboo (*Dendrocalamus strictus*). In: *VII International Congress on Plant Tissue and Cell Culture.* Amsterdam, June 24–29.

Chapter 6

Macro- and Microvegetative Propagation as a Tool in Tree Breeding as Demonstrated by Case Studies of Aspen and Cashew and Regulations for Marketing Clonal Material

Hans-J. Muhs

Federal Research Centre for Forestry and Forest Products, Institute of Forest Genetics and Forest Tree Breeding, Sieker Landstr. 2, D-2070 Grosshansdorf Germany

---

In most research fields in tree breeding, sexual or vegetative methods for propagation are used. Sexual reproduction is usually called the natural way of propagation in forest trees, while propagation by vegetative means is considered to be artificial. A number of species reproduce naturally both sexually and vegetatively, although sexual reproduction seems to be the most important. Sexual reproduction is essential for producing genetic variation, which is balanced by selection (natural and artificial) to keep the population adaptable. Vegetative reproduction is used to maintain the genotype best adapted to the site where the tree is growing (in nature), is the only way of reproduction on marginal sites (for example at high altitudes), or is used for multiplication of desirable trees which are selected artificially according to special traits. If there is a choice between sexual and vegetative propagation, the breeder should take careful account of all appropriate factors.

In addition to the natural ways of vegetative propagation (for instance root suckering, groundlayering and rooting of branches after damage to the trunk) breeders have developed many different methods for vegetative propagation, not all of which can be applied to every tree species. The reasons for this are the differences between species, the genetics and the differentiation process of the tree, and developmental and physiological stages, which are all endogenous, and in addition a number of exogenous factors. We can easily control the exogenous factors, if we know their importance and influence on propagation. But up to now we have hardly understood the influence of the endogenous factors, which therefore we do not control.

Two groups of vegetative propagation methods are used, heterovegetative (e.g. grafting) and autovegetative methods (e.g. rooting of cuttings). The principle differences between the two methods are the wide range of applications and the problem of incompatibility with the heterovegetative methods compared with autovegetative ones, where the problem of rooting may be important. Each type of method has its advantages and disadvantages, which are not discussed here. In the following discussion only two autovegetative propagation methods are referred to, which can be grouped into macro and micro methods, the latter is largely synonymous with *in vitro* methods.

The aims of vegetative propagation are:

1. maximizing genetic gain, as a last step in a breeding programme;
2. mass multiplication of selected or superior trees, the superiority of which should be selected or tested;
3. the propagation of trees for conservation.

The last aim, though important, will not be discussed here. [Recently the national concept for the conservation of forest gene resources of the Federal Republic of Germany has been published (Melchior *et al.* 1989)].

The two following case studies show the potential of methods for vegetative propagation applied to recalcitrant tree species.

Case studies

Aspen (*Populus tremula, P. tremuloides* and their hybrids)

Unlike poplars (black poplars and balsam poplars) aspen does not form root primordia in the stem and branches: therefore woody cuttings do not root easily. Results of induction of adventitious roots in woody cuttings are rather low (about 2%). But aspen does show a rather high ability to form root suckers after felling or heavy cut-back of the crown or roots. This ability may be explained by the ecophysiology of the natural habitat of the species which is mainly characterized by dry and poor soils. The branches and twigs, which may stick into the ground if a tree is blown over, do not have good conditions for rooting and therefore would not survive. In this case root suckering is shown to be much more successful and even competitive.

This is particularly obvious on fire sites and burned areas. Poplars have a quite different habitat, for instance rich and well drained soils alongside riverbanks, where rooting of branches and twigs seems to be more effective.

The method first used for vegetative propagation of aspen was the induction of root suckers and their transplantation in the field. Because of the limited numbers of transplants the method was improved by the use of root cuttings as they are able to form shoots. The next step in the improvement was the use of green (herbaceous) cuttings derived from sprouts of root cuttings; green cuttings have a potential to form roots as long as they are not lignified. (It is not known whether the lignification process itself or some other developmental or physiological changes are responsible for the reduction in rooting ability.) Up to now green cuttings from root cuttings have been used.

In our Institute this technique has been used to multiply seedlings at a very early stage (5–10 cm height), and this can be repeated several times using the non-lignified shoots coming from the axillary or adventitious buds. None of these methods is satisfactory if mass multiplication of tested superior trees is required. Therefore, in our Institute Ahuja (1986) developed a microvegetative propagation method, which resulted in a four-step method: (1) bud break and conditioning; (2) growth and proliferation of microshoots; (3) rooting; and (4) transplantation into a soil-free substrate using three different media (Ahuja, 1983). He has succeeded in culturing and rooting herbaceous cuttings derived from buds of young and adult aspen trees. Later the four-step method was simplified by combining the conditioning of the bud and the proliferation of multiple shoots from a bud in one medium, which resulted in a three-step method. Besides saving manpower and costs the elimination of one medium shortens the phase in which the organ cultures are unrooted. A further reduction from three to two steps was made possible by combining the steps of rooting and transplanting to a substrate (Ahuja, 1984). The advantage of this is the development of a better root system, which can develop in a substrate containing oxygen as in common soil. So the vitrification and etiolation of the roots, that often occurs in media can be avoided. Transplanting of rooted plants to substrate or soil is not necessary. Thus the critical, sensitive transplanting phase can be eliminated and the microroots can develop undisturbed.

The use of buds from mature trees and culture with the two-step method after Ahuja (1986, 1987) have the advantages of:

1. being able to employ the method for material from superior trees, which have been previously tested;
2. inducing organogenesis directly;
3. avoiding an extensive callus formation;
4. shortening the (obligatory) culture phase;
5. improving the economics.

From the above we may conclude that it is worthwhile to have information about naturally occurring vegetative propagation of the species and about the regeneration ability and ecophysiological responses before starting to develop a method for vegetative propagation. We may further conclude that macrotechniques and microtechniques can be combined successfully.

Cashew (*Anacardium occidentale* L.)

As far as is known cashew does not reproduce vegetatively naturally. There may be an exception, for example by ground layering, reported by Ohler (1979). Lateral branches, which touch the ground, may form roots. But it is not known whether ground layering has any significance for the strategy of reproduction of the species. Cashew has the ability to sprout (either from dormant or adventitious buds) after the crown has been damaged, mechanically or by ground fires.

Artificially cashew has been propagated vegetatively by various kinds of grafting and by air layering. Both methods seem to be feasible but costly and have their specific problems (Ohler, 1979). Now the question arises, whether rooting of cuttings could be a way to overcome the problems of incompatibility and breaks at the position of grafting and the formation of a badly developed root system if air layering is applied. So far only a few efforts have been made to develop a method for rooting cuttings, although Ohler (1979) reported the first results he had obtained.

Our own experiments with cuttings in a greenhouse showed some promising results but also a number of failures which provide evidence that cashew belongs to the group of recalcitrant tree species with respect to vegetative propagation. The age of the donor plant, from which the cuttings are derived, is important. Cuttings from young seedlings have the potential to root easily and with high success. When the leader was 10–20 cm above the cotyledons, it was cut just above the cotyledons and transferred to a bin containing peat–sand substrate without treatment with growth regulators. The leaves (usually not more than two or three) were cut in half to reduce transpiration. The bin was covered with transparent plastic to keep the humidity as high as possible. The cuttings were frequently irrigated and the water surplus was drained to avoid a standing water-table. The rooting after about 8 weeks is 90–95%. Using this simple procedure it was possible to root cuttings from hundreds of seedlings without a genotypic effect with respect to rooting ability being observed. Because of the limitation of the plant material analysed we may not have been able to observe such effects. After recovery of the decapitated seedlings, which usually produce two new shoots from axillary buds, both shoots were cut and rooted according to the procedure above. The same can be done after a further recovery of the seedling and also with the first

Fig. 6.1 Woody cuttings of a 2-year-old cashew tree in a container containing a peat/perlite substrate. Flushing of adventitious buds is frequent after some 2–3 weeks, but root formation is rare even after 6–8 weeks. After 5–6 weeks the young shoots begin to suffer because of the lack of root formation and most die within a further 3 weeks.

cuttings and cuttings from the second cycle, if they have established themselves and produced vigorous shoots of adequate size. The rootability decreased after the third propagation cycle, but was still about 40–60%. We have been able to root about five to ten cuttings from each seedling. (This number can probably be increased by improvement of the techniques.)

The easy rooting of cuttings derived from young seedlings makes possible a combined breeding and plant production programme, which may consist of the following steps:

1. selection of superior phenotypes of cashew trees ('plus' trees);
2. collection of seed from each 'plus' tree;
3. testing of progenies of all 'plus' trees;
4. parallel to this, all 'plus' trees should be grafted and used for:
    (a) the establishment of a seed orchard containing all 'plus' trees in a randomized design;
    (b) seed collection in the seed orchard after sufficient fruit setting;

Fig. 6.2. A cashew seedling, which has been decapitated. The top shoot was used for rooting (see Fig. 6.3). Young shoots from axillary buds grow up rapidly and can be also harvested for rooting.

(c) raising of seedlings and multiplication by the method for vegetative propagation described above;
(d) field planting of rooted cuttings after they have reached a certain size;
5. if results from the progeny tests (see step no. 3) are available, the seed orchard can be thinned by elimination of those 'plus' trees which produce inferior progenies.

This combined breeding and production programme results in continuous production of plants about 3 years after establishing the seed orchard. All nuts collected in the seed orchard can be used to raise seedlings and to produce rooted cuttings. Using this method plant production can be multiplied five to ten times (see step 4c). The breeding programme contains selection at two levels, first selection of superior phenotypes ('plus' trees), second the elimination of inferior trees from the seed orchard as a result of progeny tests. The improvement is dependent on different factors, such as base population, selection intensity, number of plus trees selected (100 should be minimum number), removal of inferior trees (keep at least 40 to 50 plus trees in the orchard). The number of

Fig. 6.3. A rooted cutting derived from a young seedling (see Fig. 6.2). The initiation and development of roots and new leaves occur simultaneously within 2–4 weeks.

'plus' trees given above are minimum numbers used in breeding programmes of most forest trees. They need not necessarily be optimal for cashew, but this is not yet known.

When this procedure was used with woody cuttings of older trees (1 year and older), the rooting success decreased to less than 2%. The time for rooting increased to 12 weeks or more. Thus the induction of root primordia seems to be the limiting factor. Although flushing of dormant or adventitious buds occurred frequently the young shoots were able to grow up to 10 cm in length, but they died after some weeks if no roots were formed. We may conclude that woody cuttings have lost their potential for easy rooting, although it is not known whether it is lignification or something else which affects the rooting ability. Investigations are under way to improve the rooting of woody and semilignified cuttings by varying various factors, such as pretreatment with growth regulators, light intensity including darkness, hydroponic system versus substrate as rooting media.

Micropropagation of cashew is not yet applicable for mass propagation, although there has been some success in regeneration of plantlets by tissue culture (Lievens *et al.*, 1989). The authors concluded that the presence of phenolic compounds make it difficult to establish a culture. There are differences between clones.

The main problems we are facing in vegetative propagation of cashew are the phenolic compounds, the very long initiation phase for roots, the lack of knowledge of the dormancy periods and their regulation during a season, of variations between different branches of a tree, and of the sensitivity to transplanting. In many respects there are similarities between cashew and the central European oaks if one compares the rootability of cuttings and the regeneration of plantlets by micropropagation. This led to a test of all the propagation techniques and treatments which have been successful in propagating oaks vegetatively (Spethmann, 1986). The aim is to develop a method which is easy to handle even in tropical countries without using highly sophisticated techniques and expensive chemicals. Basic knowledge of shoot and root growth and the physiology of ageing are required. Starting from tissue originating from young plants (e.g. seedlings) should be preferred, as long as rejuvenation of mature tissue is not or hardly possible. The effects of topophysis and cyclophysis can be very strong and visible after years of cultivation. It may even be that the root system of rooted cuttings derived from young seedlings differs from that of rooted cuttings derived from mature trees. Generally the former tend to develop a tap root more frequently, while the latter do not. Thus it has to be analysed whether the plants produced by cuttings are more viable and even more vigorous than seedlings. This seems to be the case in some species like black and balsam poplar (*Populus* spp.) and also in Norway spruce (*Pices abies* (L.) Karst), but it cannot be generalized. On the other hand different propagation techniques may have an impact on cultivation and the growth potential of the plants produced compared with seedlings. It is well known that in many tree species plantlets produced via tissue culture have reduced growth potential and for a long time remain inferior to seedlings. This can be overcome by improving the propagation techniques in many cases.

Impacts of using clonal material at a large scale

The use of clones in forestry has various advantages over natural seedlings, which can be summarized as follows:

1. high yield;
2. uniform wood quality;
3. high percentage of marketable timber;
4. low production costs because of easy silvicultural treatments and harvest;
5. shorter rotation period.

The impacts of the environment on the cultivar are difficult to analyse because of the long time required for impacts to become visible or measurable. Cultivars of some forest tree species including hybrids like poplars (*Populus* spp.) in Europe and North America or sugi (*Cryptomeria japonica*) in Japan, which have been vegetatively propagated for more than 100 years in large areas, do not show serious signs of destabilization, although there are changes in vigour and in resistance to diseases or pests. Changes in vigour can be explained by an accumulation of viruses in the tissue, which do not appear to cause disease but reduce the vigour. Changes in resistance are mostly due to changes in virulence or pathogenicity of the parasite population. If the loss of rootability or decrease of growth vigour is observed, then ageing may be a cause; this can be overcome by rejuvenation of the tissue before starting a new propagation cycle. In species where there are difficulties in rejuvenating the tissue, an accumulation and manifestation of age-dependent characteristics occurs in the plant material. Cultivars of these species cannot be propagated interminably. But the integrity of the genome and germplasm of the cultivars seems to be rather stable after many propagation cycles within a period of over 100 years. Mutations do occur within this time span, but they affect the viability of the cultivar only in exceptional cases. Formerly it was assumed that mutations were the main source of the destabilization of the cultivar.

Based on our experience we conclude that environmental factors do not have a destabilizing effect *per se* on vegetatively propagated cultivars. So far we have been using the term 'cultivar' for vegetatively propagated plant material. This is a historical term, instead we will now replace it by the term 'clone'. A clone consists of an ortet, the original specimen, and the ramets derived by vegetative propagation from the ortet. A clone is identified by pedigree, whereas a specimen of a cultivar is not, because the 'ortet' remains unknown.

The impacts of clones on the environment are different from those of cultivars. Impacts can be divided into two groups:

1. impacts connected with the replacement of natural ecosystems by mono- or multiclonal plantations;
2. impacts connected with the spread of certain genotypes, which have been selected for clonal plantations, and also with the increase of their relative frequency within a stand or even within a region.

These impacts are not independent but they are causes for different concerns. The *first* group leads to a change of the environment by reducing the area of natural habitats or ecosystems. The result can be a narrowing of the range of the habitat variables or shifting of those variables towards a marginal system. In other words, the richer soil types will be 'cultivated' with clones, while the poorer soil types remain as relicts for the natural ecosystem. In some cases this results in a loss of certain species which are not able to adapt to the changing habitat. If the species being replaced and the species of the clonal planting stock are the same, there will be a considerable loss of genetic variability, although locally adapted populations tend to survive. In the long run this can be detrimental to the species and its evolution. Man can (and should) help threatened species by conservation of genetic resources *in situ* or in serious cases by evacuation and storing germplasm *ex situ.*

Impacts of the *second* group are also of concern even if there is no replacement of a natural ecosystem, for instance the afforestation of abandoned agricultural land by clonal plantations. The concern in this case is the multiple use of a few unique clones, which leads to a high degree of uniformity within a stand. This may bring a high risk of loss in case of diseases or pests and a less flexible response of the artificial population to various biotic and abiotic factors. If the clonal plantation is used as a seed stand for natural regeneration at the end of the rotation period, the seeds produced will have a narrow genetic base with a high probability of *de facto*-selfed inbreeding and may be less adapted to local conditions. Pathogens may find a good field for action within the homogeneous clonal populations and may even affect local natural ecosystems more seriously than would be the case without such 'breeding fields'. The only way to avoid these problems is to reduce the number of ramets per clone or to make multiclonal mixtures that have as much genetic variability as seedling populations.

Regulations for the marketing of forest reproductive material

The reason for the introduction of regulations for the marketing of forest reproductive material at an international level was not to minimize detrimental impacts on the environment of using cultivars and clones, but to protect the customers' interests, although the first argument can be expected to gain more and more importance in discussions about regulations for the use of clonal material in forestry. The historical roots, however, go back to 1925, when in Germany and some other Central European countries efforts were initiated to regulate the use of material for planting in forests. At that time the international seed trade was expanding and it was realized that there should be rules to protect forest regeneration from unsuitable sources of seed. Since then the rules have been changed several times and attained legal status in 1934. Since the Second World War two international regulations have been set up, the EEC-Directives and the OECD Scheme, which will be briefly described below. A more detailed treatise on this matter is given elsewhere (Muhs, 1991).

International regulations

The European Economic Community (EEC) enacted its Directives on marketing of forest reproductive material in 1966 and supplemented and

amended the regulations in 1969 and 1975. These Directives are manda-
tory for all EEC-member countries and must be incorporated into national
law within a given time. The Organization for Economic Co-operation and
Development (OECD) published in 1974 the Scheme for the control of
forest reproductive material circulating in international trade. The scheme
is open to all countries including EEC-member countries. Although both
international regulations are based on the same principles, such as the
principle of approval, the principal of identification and the principle of
control, there are some differences (Muhs, 1991). The OECD-Scheme has
four categories:

1. source identified reproductive material (yellow label);
2. selected reproductive material (green label);
3. untested seed orchard material (pink label);
4. tested reproductive material (blue label).

The EEC-Directives accept only two categories:

1. selected;
2. tested reproductive material.

Both regulations include sexually reproduced material (seed, seedlings) and
vegetatively propagated material (for instance rooted cuttings). At the time
when the regulations were adopted clonal forestry was restricted to some
species of local interest or of minor importance like poplars and willows. In
the meantime, however, techniques have been developed for vegetative
propagation, using both cuttings and tissue culture techniques. Now that
many forest tree species, like Norway spruce, can be propagated vegetatively on a large scale regulations are needed for the marketing of clonally
produced plant material. Both international organizations are preparing to
amend their rules in order to incorporate regulations for the approval,
identification and control of clonal planting stock.

National regulations for the marketing of clones and clonal mixtures

So far only two countries, Germany and Sweden, have enacted laws for the
release of clones and clonal mixtures to the market (Table 6.1). Both laws
are based on the approval of basic material and govern the marketing of
reproductive material derived from the basic material. They both include
aspects of minimization of detrimental effects of using clones and clonal
mixtures. In addition to these laws, rules which govern the deployment,
rather than the approval and marketing of clonal planting stock, have been
established or are being considered in some other countries. These rules are
only of regional importance and will not be considered further here. In Table
6.1 some striking features of the laws from both Sweden and Germany are
outlined. A more detailed description is given elsewhere (Muhs, 1991).

The synopsis given in Table 6.1, shows the different approaches to the solution of setting up adequate rules for marketing of clones and multi-clonal mixtures. The most critical item in these regulations is the fixing of the minimum number of clones per mixture. Libby (1988) among other authors has dealt with this matter and found three general guidelines:

1. a mixture of large numbers of clones (100 and more) is about as safe as a similar mixture of genetically diverse seedlings;
2. mixtures or regional deployment of very small numbers of clones is not safe, and commitment to two to four clones is often worse than monoclonal plantations; and
3. regional deployment of modest numbers ranging from 7–99 clones is about as safe as deployment of large numbers of clones, and offers substantial advantage as well (Libby and Rauter, 1984).

Another critical item, which has not been considered in the German and Swedish laws so far, is the influence of the propagation technique on the viability and growth vigour of the plant material. It is known that plants derived from rooting of cuttings and plantlets derived by *in vitro* techniques can behave differently in the field, showing considerable differences in survival and growth vigour even after several years: this could be a subject for regulation.

Conclusions

As the techniques for vegetative propagation are developing rapidly, it is necessary to include guidelines for the marketing of clonally produced plant material in the international regulations (OECD-Scheme and EEC Directives). The testing of clones as well as the proof of the propagation techniques used for the production of clonal planting stock should be a must in a breeding programme. Before giving permission for the release of clones and clonal mixtures the impacts of the use of such plant material should be studied. Also the alternatives to the use of clonal planting stock should be carefully investigated and the following questions answered:

1. Is the root system of the clonally propagated plant equivalent or superior to that of seedlings of the same species?
2. Is the rootability of a clone a selection criterion? If it is, due account should be taken of the fact that selection for easy rootability may affect the genetic gain or reduce the genetic diversity of the clonal mixture.
3. Can the stock material be kept juvenile for the time of propagation and what methods for rejuvenation are available? The accumulation of the effects of topophysis and cyclophysis or ageing in the stock material should be avoided.

Table 6.1. Synopsis of laws governing the marketing of clonal material.

	Sweden	Germany

1. *Legal status of the rules:*

	law since 1982, amended in 1991	law since 1979, details are laid down in a General Administrative Regulation amended in 1985

2. *Use of clonal material as single clones or clonal mixtures:*

	multiclonal mixtures are recommended, row planting is preferred over block planting, if clones are kept identifiable	multiclonal mixtures only

3. *Species covered other than poplars:*

	conifers	all species (total 18) under the law, for the genus *Populus* single clones and mixtures are allowed

4. *Use of untested clones:*

	allowed	not allowed

5. *Use of bulk propagated plant material:*

	allowed: up to 200 ramets from each plant derived from seed or from an unidentified clone	not allowed

6. *Testing procedure, duration of test:*

	at test level 1 the clones have to be tested at two sites for 6 years, at test level 2 the test at level 1 should be prolonged for 3 further years (total 9 years) additionally two new replications of the whole test at different sites have to be included at test level 2	tests have to be established at three sites at least, the duration of the tests depends on the species and is not fixed, usually 10–30 years are commonly used in progeny testing which is recommended for clonal testing also

7. *Minimum numbers of clones to be used in a clonal mixture:*

	dependent on the test level. The percentage for a single clone in a plantation should – not exceed 1.5% for untested clones (this corresponds to a minimum number of 67 clones),	dependent on the species. For main forest tree species, like Norway spruce, the minimum number of clones is 500. If clones for special site conditions are tested this number can be reduced to 100. For less important species the

Table 6.1. Continued

Sweden	Germany
– not exceed 2.5% at test level 1 (this corresponds to a minimum number of 40 clones), – not exceed 3.5% at test level 2 (this corresponds to a minimum number of 29 clones)	minimum number is 100. If clones for special site conditions are tested this number can be reduced to 20

8. *Limitation of approval:*

dependent on the test level. For untested clones not more than 50,000, for clones of test level 1 not more than 250,000, for clones of test level 2 not more than 700,000 ramets shall be produced. This number does not include the numbers at lower levels. If quota at lower levels are not used they can be added at higher levels	the approval is limited to 10 years, prolongation for another 10 years is possible

Clonal forestry is a different way of using forest land resources compared with the management of a natural or traditional forest system. It offers advantages, if we take account of the principle of sustained yield over more than just one rotation period.

References

Ahuja, M.R. (1983) Somatic cell genetics and rapid clonal propagation of aspen. *Silvae Genetica* 32, 131–5.

Ahuja, M.R. (1984) A commercially feasible micropropagation method for aspen. *Silvae Genetica* 33, 174–6.

Ahuja, M.R. (1986) Aspen. In: Evans, D.A., Sharp, W.R. & Ammirato, P.J. (eds), *Handbook of Plant Cell Culture.* Macmillan Publishing Company, New York, pp. 626–51.

Ahuja, M.R. (1987) *In vitro* propagation of poplar and aspen. In: Bonga, J.M. & Durzan, D.J. (eds), *Cell and Tissue Culture in Forestry*, Vol. 3. Martinus Nijhoff Publishers, Dordrecht, pp. 207–23.

Libby, W.J. (1988) Testing and deployment of brave new plantlets. In: Ahuja, M.R.

(ed.), *Somatic Cell Genetics of Woody Plants.* Kluwer Academic Publishers, Dordrecht, pp. 201–9.

Libby, W.J. & Rauter, R.M. (1984) Advantages of clonal forestry. *The Forestry Chronicle* 60, 145–9.

Lievens, Ch., Pylyser, M. & Boxus, Ph. (1989) Clonal propagation of *Anacardium occidentale* by tissue culture. *Fruits* 44, 10.

Melchior, G.H., Becker, A., Behm, A., Dörflinger, H., Franke, A., Kleinschmit, J., Muhs, H.-J., Schmitt, H.-P., Stephan, B.R., Tabel, U., Weisgerber, H. & Widmaier, Th. (1989) Konzept zur Erhaltung forstlicher Genressourcen in der Bundesrepublik Deutschland. *Forst und Holz* 44 (15), 379–404.

Muhs, H.-J. (1991) Policies, regulations and laws affecting clonal forestry. In: Ahuja, M.R. & Libby, W.J. (eds), *Clonal Forestry: Genetics, Biotechnology and Application.* Springer Verlag, Heidelberg.

Ohler, J.G. (1979) Cashew. *Communication 71, Department of Agricultural Research, Koninklijk Instituut voor de Tropen,* Amsterdam, The Netherlands, pp. 260.

Spethmann, W. (1986) Stecklingsvermehrung von Stiel- und Traubeneiche (*Quercus robur* L. and *Quercus petraea* (Matt.) Liebl.). *Schriften aus der Forstlichen Fakultät der Universität Göttingen und der Niedersächsischen Forstlichen Versuchsanstalt,* J.D. Sauerländers Verlag, Frankfurt/Main, 86, pp. 99.

Chapter 7

Experiences in Vegetative Propagation of *Populus* and *Salix* and Problems Related to Clonal Strategies

Louis Zsuffa

Faculty of Forestry, University of Toronto,
33 Willcocks Street, Toronto, Ontario, M5S 3B3 Canada

Poplars (species and interspecific hybrids of sections *Aigeiros* Duby, and *Tacamahaca* Spach.), aspens (*Leuce* Duby. poplars) and willows (*Salix* L.) exhibit a variety of morphological, anatomical and physiological differences. Due to their fast growth and the fact that they may be planted on many varied sites both in and out of forests, poplars, aspens and willows offer ample possibilities, especially in the northern temperate zone, of opening up additional sources of raw material and thus helping to meet the continually increasing demand for wood. For example, biomass plantations of willow clones planted at a density of about 20,000 cuttings per hectare, cultivated by agricultural methods, and harvested in 2–4 year cycles, produced 12–15 oven-dry t ha^{-1} y^{-1} of stem wood in practical scale trials. In intensively cultivated field trials, biomass yields have been reported exceeding 30 oven-dry t ha^{-1} y^{-1}. Furthermore, these fast-growing trees can enrich the forests of northern regions which are often lacking in tree variety, and give a new appearance to parts of the open landscape where there are few trees and bushes.

The advantages of vegetative propagation for poplars and willows have been known for a long time. By means of the conventional methods of rooting of cuttings poplars and willows are reproduced clonally in large numbers each year in many countries and for practical cultivation. However, this method of reproduction cannot be used in the same way for all species. Difficulties occur particularly when no or only few preformed or latent root primordia exist in the shoot. Thus it is more difficult to achieve practical large-scale propagation by means of cuttings for the aspens. Only recent experiences using *in vitro* techniques show positive results.

Clonal propagation

One of the most important characteristics of many poplars and willows is their readiness to propagate vegetatively by rooting of stem cuttings. The advantage of this for practical cultivation is obvious. Besides the low cost of producing new plants the possibility of achieving genetically identical material by very simple means is very tempting. Silviculturally, stem cuttings are used for propagating clones not only in the nursery but also in the field, where unrooted cuttings are often preferred over rooted cuttings as an inexpensive and easy-to-handle planting stock.

Factors affecting rooting of cuttings are either internal or environmental, or result from the interaction of the two (Zsuffa, 1976). Internal factors influencing rooting are related to the genetic, morphological and physiological characteristics of the ortet and ramet. The most significant environmental influences on rooting are light, temperature and moisture. Genotype–environmental interactions cause specific changes in the biology and chemistry of the ortet and ramet, which affect the rooting.

The propagation ability of willows can be demonstrated by the method described by Ericsson (1982), where it is the human factor, more than the plants, that limits the speed of propagation. Through enhanced axillary shoot formation of the ortet, more than 10^9 ramets could be produced in 1 year under glasshouse conditions. This method offers a very high rate of multiplication, and requires a smaller input of labour and technical facilities than tissue culture techniques (Fig. 7.1). A similar technique has been employed with both aspen and cottonwood type *Populus* for establishing research plantations (Faltonson *et al.*, 1983).

The rooting ability of some poplar, aspen and willow species, which show variable or poor rooting from stem cuttings, can be improved by crossing with a related species which shows better rooting. Thus, most poplar species of *Aigeiros* and *Tacamahaca* sections root with ease. However, the rooting ability of *Populus deltoides* varies and sometimes, under unfavourable field conditions, large portions of its cuttings perish. Cuttings of *P. nigra* and *P. balsamifera* Duroi root more readily. These species transmit their better rooting ability to hybrids with *P. deltoides*. In fact, cuttings of *P. × euramericana* and *P. × jackii* Sarg. (*P. deltoides × balsamifera*) clones commonly root even under adverse field conditions.

Within the poplars of the *Leuce* section, the European white poplar (*P. alba*) is the only species which can be propagated with dormant stem cuttings. White poplar creates fertile and vigorously growing hybrids with aspen species and at the same time transmits its rooting ability to the hybrids, as reported, for example by Heimburger (1968), for crosses of *P. alba* with *P. grandidentata* Michx., *P. tremuloides*, *P. tremula*, *P. davidiana* (Dode) Schneid. and *P. sieboldii* Miq. By backcrossing the hybrids to *P. alba* and by producing multiple hybrids of unrelated parents,

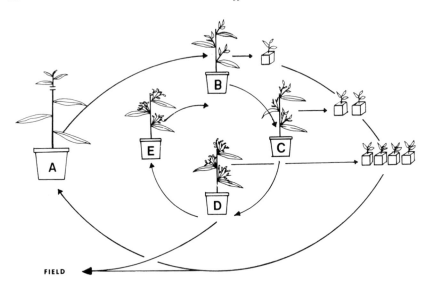

Fig. 7.1. Sequential steps during cutting production from axillary shoots in *Salix*. After removing the shoot apex (A) axillary shoots may be harvested repeatedly (B–E) for long periods. The rate of propagation is accelerated if the rooted cuttings are used as stock plants for a number of propagation cycles (from Ericsson, 1982).

such as *P. canescens* × (*alba* × *grandidentata*), *P.* (*alba* × *davidiana*) × (*alba* × *grandidentata*), heterogeneous progenies with outstanding individuals were produced, in which the rooting ability was strengthened (Heimburger, 1968; Zufa, 1969).

The poor rooting ability of aspens is a major setback in cloning and operational use of genetically improved stock. Most attempts to improve the rooting of cuttings have failed, especially with *P. tremuloides*, *P. tremula*, and related aspens. Therefore in recent years new ways for solving these problems were attempted by the use of *in vitro* techniques. Bonga and Durzan (1982) provide comprehensive information on the present state of knowledge and the importance of this technique for forestry. Research to date has concentrated mainly on *Populus tremuloides* and *P. tremula* for hybrids of *P. alba* with *P. glandulosa* or *P. grandidentata* (Ahuja, 1983; Fröhlich and Weisgerber, 1987).

Poplars of section *Leuce* can be propagated by planting root segments. Cuttings approximately 10 cm in length and 0.5–3 cm in diameter are made from dormant root systems. When planted in nursery beds, each segment will give rise to one or more new plants just as new sprouts form from the roots when natural stands are cut or burned. This propagation technique has been employed in the production of ornamental aspen clones and is now being studied in combination with tissue culture techniques to try to find a rapid, economic clonal deployment strategy for the hybrid aspen clones (Chun and Hall, 1984).

Clonal testing

Clonal forestry is only as reliable as the clones used for plantings. The reliability of clones depends on selection criteria and testing. Improper choice and application of criteria may cause difficulties in clonal selection.

Criteria often used for selection of poplar and willow clones include: good rooting ability; resistance to diseases; desired wood quality; stem form and branching habit; positive reaction to site management; and regeneration by coppicing. Clones can be directly observed for some of these criteria, such as for rooting ability, stem form and branching habit, and this facilitates the selection. The process becomes more complicated and less reliable, when only indirect evidence can be provided for traits, such as when resistance to disease is judged on the basis of presence or absence of some biochemical compounds, and growth on the basis of photosynthesis. The reliability of such criteria depends largely on the strength of correlation between the trait and the indirect criterion judged: if the correlation is strong, such methods may even result in significant savings in time and amount of work.

Difficulties in clonal selection are encountered because of the time element and the effect of environment (Zsuffa, 1975). We are pressed by time to judge the trees on the basis of early performance. Some important clonal traits, such as rooting ability and stem form, can be detected at a juvenile age. However, other significant traits, such as growth rate and disease resistance, may change with age. For example, in a clonal trial in Ontario, Canada, the ranking of clones changed considerably during the first 6 years of growth (Table 7.1). This meant that several years of observation were needed for proper clonal selection. Also, the fast starter clones would be better suited for 1–3 year harvesting cycles in biomass plantations, whereas longer rotations might require different clones (Zsuffa, 1975).

Many important clonal characteristics are largely influenced by environment. Such is the growth intensity, which depends significantly on the site. Various poplar and willow clones vary in their site requirements, often even if they are of the same species origin. In euramerican poplar clonal trials across a range of sites, some clones grew relatively well on a number of sites, whereas some performed extremely well on some sites but showed poor growth on others (Zufa and Zivanov, 1966).

Clonal tests are usually organized at several levels, from the initial screening of single copy genotypes (in progeny trials) through clonal performance testing. Libby (1987) recommends testing for clonal forestry at four levels.

Table 7.1. Ranking of clones according to mean heights at Huronia District, Flos Twp. (southern Ontario, lat. 44°10'N, long. 80°W) clonal test established in the spring of 1969, on Tioga loamy sand, at 10 × 10ft (3 m × 3 m) spacing. All clones in the test are of *Populus* × *euramericana* (Dode) Guinier (*P. deltoides* Bartr. × *P. nigra* L.) origin. The solid and broken lines indicate the changes in ranking of some clones.

Clones	Ranking at age (years)				
	1	2	3	4	6
I-10/56	1 (1.7m)*	10 (3.7m)	10 (4.6m)	4 (6.1m)	11 (8.5m)
I-61/59	2	6	4	11	20
Negrito de Granada	3	3	11	5	5
I-214	4	5	8	10	4
Ostia	5	8	1	20	15
Jacometti 78B	6	12	20	8	8
B-56	7	4	5	1	22
I-262	8	1	13	12	10
I-55/56	9	7	12	15	12
Eugenii[†]	10	9	3	13	14
Regenere batard d'Hauterive	11	11	18	19	1
Canada blank	12	13	6	3	19
Regenerata Harff	13	15	9	6	13
I-132/56	14	19	7	18	6
Blank de Poitou	15	18	22	9	24
Virginie de Frignicourt	16	20	2	22	3
Chopa de Santa Fé	17	2	15	2	18
Virginie de Nancy	18	14	21	7	2
Dromling	19	22	14	14	21
Robusta	20	21	19	21	9
Regenere d'Aube	21	16	16	24	7
I-45/51	22	23	24	17	23
I-455	23	24	23	16	17
Tardif de Champagne	24 (1m)	17 (2.5m)	17 (3.1m)	23 (4.4m)	16 (5.9m)

*Mean height
†Clone developed from a Carolina poplar tree, selected in Orono nursery

Level 1 – Initial screening, involves seedlings, although in some cases the plants screened will be single copy plantlets of tissue culture or embryoid origin.

Level 2 – Candidacy testing, with large numbers of clones and a small number of ramets per clone.

Level 3 – Clonal performance testing – more extensive testing of better candidate clones. The number of clones at this level of testing will be moderately small, and the number of ramets per clone should be large.

Level 4 – Compatibility trials attempt to identify compatible sets of clones that can be advantageously grown in sequenced mixtures. This level of testing will be done by advanced programmes, with a relatively small number of highly successful and well known clones.

Many programmes practise the first three levels, while level 4 has rarely been used in clonal testing of poplars, aspens and willows.

Clonal plantation strategies

Monoclonal planting is common in all forms of poplar and willow plantations (Zsuffa, 1985). The reasons for, and practicality of, monoclonal poplar and willow culture are: ease of propagation by rooting of cuttings; the advantages of growing and utilizing plantations of uniform trees; large variation within populations and families; and the occurrence of unusual, desired individual types. The ease of vegetative propagation alone, if dealing with relatively uniform populations and sibs containing many desired types, could have led to multiclonal cultures. However, poplar species in general, and families of interspecific hybrid poplars in particular, are well-noted for large tree-to-tree variation. In fact *P. nigra* clonal selections of fastigiate forms, such as cvs. *italica, thevestina* and *plantierensis*, which have been propagated for centuries, represent unique and rare occurrences. Old selections of euramerican poplars, such as cvs. *serotina, regenerata* and *robusta*, are also distinguished by particular features of form, growth and site adaptation. For this reason, and because of the relatively late start of planned poplar and willow breeding programmes, the number of clones used in monoclonal plantations throughout the world has remained relatively small.

The International Poplar Commission has registered 52 clones and has eight more clones which are candidates for registration (FAO, 1983). Most of these clones are euramerican poplar hybrids. The Commission's textbook on poplars and willows (FAO, 1979) describes five clones of *P. nigra*, 22 clones of *P. deltoides* (in use since 1970), 44 clones of euramerican poplar (about half of them in use since 1970), and 20 clones of balsam poplars and their hybrids (eight of which have been in use since 1970). Thus, the total number of poplar clones registered and in commercial use is fewer than 100.

The situation with willow clones is similar in that the total number of clones in commercial use does not exceed 100. However, many of these are osier willows (shrubs); few clones are actual trees.

While osier and biomass (energy) plantations are always monoclonal, tree plantations of willows, especially of *Salix alba* L. along the Danube River, are occasionally multiclonal. This is because special selections were not made earlier within this species which is native to the flood plain of the river, and the plantations established on similar sites are extensively managed as forest stands.

Only a portion of registered poplar and willow clones are widely used in plantations. According to the reports of the National Poplar Commissions submitted to the 16th Session of the International Poplar Commission, the number of clones planted on a large scale is very small in most of the countries (Chardenon, 1980). In Hungary, three clones (*P.* × *euramericana* cvs. *Robusta*, I-214, and *Marilandica*) represent 81.0% of the plantations; in the Netherlands, euramerican poplar clones *Robusta*, and *Zealand* account for 60%; and in France, *Robusta* and I-214 account for 70% of the plantings. In the Republic of Korea, euramerican poplar clones I-214 and I-476 are planted almost entirely on the plains, while seedlings of a single hybrid aspen family (*P. alba* × *glandulosa*) are planted in the hills and lower mountain ranges on a total of 428000 ha. A few of the clones represent well over 50% of all poplar plantations, the extent of which can be estimated at more than 1500000 ha worldwide.

The situation in willow plantations with regard to the number of clones registered and planted over large areas is similar to poplar. In England, a single clone (*S. alba* × *coerulea*) has been planted and used for the production of cricket bats. In osiers, as well as in tree plantings, a handful of clones are favoured and used throughout the world.

The need for using more and diversified clonal stock was realized in poplar and willow culture in the 1960s in conjunction with the dawn of new ideas in plantation technology, spreading of plantations to different sites, and well-planned breeding programmes. The ideas on integrated use of trees (the biomass concept), and new needs and possibilities in biomass utilization (such as for composite boards, energy, food) led to studies of matching clonal characteristics to production and utilization technology. These studies showed significant clonal variation and resulted in new clonal selections and diversification.

After the Second World War, poplar plantations spread to new and different sites, such as forest sites, upland sites, and different countries and climatic zones. Occasionally, traditional clones did not perform satisfactorily in such conditions and a search for new selections was initiated. At the same time, some of the well-established breeding programmes have come to fruition. All this resulted in a variety of new breeding stock with new combinations of hybrids, and a large number of new clonal selections. Although the need for diversification of clonal stock was realized and breeding efforts for this diversification were made, multiclonal concepts were only considered and developed in a few cases.

According to Libby (1987), there are two major alternatives for clonal deployment: WIMPs (Widespread Intimately Mixed Plantations) and MOMS (Mosaics of Monoclonal Stands). In practice, these two main concepts are considered and applied for clonal mixtures in poplar plantations in (i) the use of mosaics of relatively small monoclonal blocks which are similar in size, and (ii) the use of row-to-row or tree-to-tree type clonal mixtures.

An alternative strategy is available for the aspens and *Salix* species that root sprout (Hall, 1982). WIMPs can be established and managed for a long enough first rotation to force substantial tree-to-tree competition. Then, after the first harvest and regeneration from sprouts the differential competitive ability of the individual clones on the different microsites will begin to change the plantation into more of a MOMS configuration. The advantage of this approach is that the actual site conditions are making the final selection of the clone(s) to be grown there.

The concept of mosaics of monoclonal blocks in poplar plantations was developed and implemented in Ontario, Canada (OMNR, 1983). According to this concept, clones are carefully matched to site, plantation system and the production goals. Sites are planted with several, rather than one or two, different clones. Monoclonal blocks are no larger than 5 ha in size. Each clone in production is continuously being tested for growth and pest resistance. As better clones become available from on-going breeding work, the poorer clones are removed from nursery propagation and planting. The number of clones in propagation and planting is more than 50 at any one time, and 5–10% of the clones in production are changed annually by phasing-out and new introductions.

The use of row-to-row and tree-to-tree clonal mixtures was attempted in the Netherlands and Germany. Kolster (1978) investigated the possibility of decreasing monoclonal risks by mixing two clones by rows in plantations established with 500–625 trees ha^{-1}. The results indicated that this method did not offer practical possibilities at these spacings. The growth of one clone was suppressed by another. Also, the loss in wood production by one clone was not compensated for by the increased growth of the trees of the other clone. For this reason, the mixing of rows of different clones was discouraged. The same result was expected when individual trees of different clones were mixed.

The concept of mixtures of individual trees of several clones, or 'multiclonal varieties', was furthered by Weisgerber (1979) in West Germany. At present, multiclonal varieties with 5–20 individual clones respectively are registered for trade in that country. Identification of the clones in the nurseries is carried out by a combined method with assessments of morphological, physiological and biochemical characteristics.

Multiclonal varieties should be of compatible sets of clones that can advantageously be grown in sequenced mixtures. Clones for such mixtures should be tested in compatibility trials (Libby, 1987), which demand a relatively long-term programme. Multiclonal varieties, in which neighbouring trees make complementary demands on the site, are theoretically more productive per unit area than monoclonal varieties, where neighbouring trees make competing demands. However, the experimental evidence in support of this theory is not strong or consistent.

Multiclonal varieties, if not thoroughly tested, can create problems of maintenance and control. In addition, it will be difficult to maintain a certain predetermined clonal mixture of the stock because of problems in clonal identification and differences in the rate of rooting of cuttings of various clones. Also, it may be difficult to judge the performance of individual clones in the mixture and adjust and improve the multiclonal variety accordingly. Consequently, the preferred concept of clonal mixtures in pure and intensively managed plantations of poplars and willows is in uniform monoclonal blocks of sizes dictated by site and other needs. In his study of models of clonal plantations, Libby (1980) came to the conclusion that monoclonal plantations are frequently the best strategy, while mixtures of two or three clones are frequently the worst and rarely, or never, the best. Mixtures of a large number of clones in either multiclonal or monoclonal plantings are as safe as seedling plantations.

Clonal strategies for the future

Multiclonal plantations are a desirable alternative to existing monoclonal plantations in poplars and willows. The concept of 'multiclonal varieties' in individual tree mixtures may be feasible for forest stands. The concept of mosaics of monoclonal blocks which are relatively small and similar in size and in which clones are matched to soils, plantation systems and production objectives appears to be a reasonable choice for intensively managed plantations.

Until recently, poplar, aspen and willow breeding programmes were unable to produce a sufficient number of clones of known and desired qualities for specific sites, plantation systems and production goals. Breeding programmes currently underway, especially in poplars, have renewed interest, and new considerations given to genetics and improvement of poplars and willows have provided hope for the realization of such goals. The genetic base for breeding has been significantly broadened, large collections of native and exotic species have been assembled and genetic variation studies are underway. As well, studies of genetic–environment interactions and parameters for many traits are yielding results. New and better planned crosses are producing rich pools of hybrids for selection and clonal development. Internationally co-ordinated programmes, such as those within the frame of the International Energy Agency (IEA), the

International Union of Forest Research Organizations (IUFRO) and the International Poplar Commission (IPC) provide good exchange of information and faster results. A broadened genetic base, new knowledge, and good co-ordination can yield the desired stock for proper multiclonal management of poplars, aspens and willows.

Pure clonal stands are easier for the silviculturalist, the nurseryman, the designated authority for control of planting stock and the industrial user. However, if the clonal balancing is left to the user, there is the danger that one or two clones will be favoured and thus constitute the majority of plantings. In order to prevent this, each single clone should have a high and specific usability. With such clones mosaics of monoclonal plantations, consisting of clones tailored to sites, plantation systems and production goals will be advantageous for intensive, industrial production. On the other hand, mixtures of clones of similar characteristics, or multiclonal varieties, may be advanageous in forest stands when components, site, or their interactions vary and/or are not properly known (Heybrook, 1978). Stock for multiclonal management has yet to be developed for most situations; however, active breeding programmes in poplars, aspens and willows have the capacity to produce such stock.

Recommendations

Any clonal strategy will be only as good as the clones used for the plantings. Diseases appear to be a major limiting factor in the use of clones, either in monoclonal or multiclonal plantations. Therefore increased research efforts in breeding resistance, pathogen control mechanisms, clonal testing and screening for resistance and disease–loss relationship are recommended.

Focusing on studies determining clonal characteristics important for matching clones to sites and production goals is also recommended. This knowledge is necessary for the multiclonal strategy using mosaics of monoclonal stands (MOMS) (Libby, 1987), and also for widespread intimately mixed plantations (WIMPs). Clonal testing for the latter in compatibility trials, as recommended by Libby (1987) should be carried out.

Continuing research efforts in genetics and breeding of poplars, aspens and willows will secure a broadening of the genetic base, the necessary knowledge for genetic manipulation and for the selection of genetic material and will secure the variety of clones needed for the implementation of any clonal strategy. Research to improve vegetative propagation, especially with aspens and those poplar and willow species which show inconsistency in rooting, is recommended.

Finally, building up national and international information systems on clones and their characteristics, furthering international co-operation in joint testing and exchange of breeding and clonal stock is important. These activities will advance the development and spread of suitable and reliable stock for future clonal plantations and strategies.

References

Ahuja, M.R. (1983) Somatic cell differentiation and rapid clonal propagation of aspen. *Silvae Genetica* 32, 131–5.

Bonga, J.M. & Durzan, D.J. (1982) *Tissue Culture in Forestry.* Den Haag, Boston and London.

Chardenon, J. (1980) Summary of national reports on questions not the subject of analysis by subsidiary bodies of the Commission. *International Poplar Commission, 16th Session, Turkey,* November 4–8, 1980. FO:CIP/80/12. 4pp. (mimeo).

Chun, Y.W. & Hall, R.B. (1984) Survival and early growth of *Populus alba* × *P. grandidentata in vitro* culture plantlets in soil. *Journal of the Korean Forestry Society* 66, 1–7.

Ericsson, T. (1982) Propagation of *Salix* through miro-cutings. Paper presented at the *North American Poplar Council 19th Annual Meeting,* July 20–22, 1982. Rhinelander, Wisconsin, USA.

Faltonson, R., Thompson, D. & Gordon, J.C. (1983) Propagation of poplar clones for controlled-environment studies. In: *Methods of Rapid, Early Selection of Poplar Clones for Maximum Yield Potential: A Manual of Procedures.* USDA Forest Service Technical Report NC-81, pp. 1–11.

FAO (1979) Poplars and willows in wood production and land use. *FAO Forestry Series* No. 10 328 p.

FAO (1983) *Report of the 31st Session of the Executive Committee of the International Poplar Commission.* Casale Monferrato, Italy, September 6–8, 1982. FO:CIP/82/Rep. 38 pp.

Fröhlich, H.J. & Weisgerber, H. (1987) Züchtungswege bis zum Aufbau von Mehrklonsorten. *Forstwissenschaftliches Centralblatt* 106, 312–28.

Hall, R.B. (1982) Breeding trees for intensive culture. *Proceedings of the IUFRO Joint Meeting of Working Parties on Genetics About Breeding Strategies Including Multiclonal Varieties.* Escherode, FRG, pp. 182–93.

Heimburger, C. (1968) Poplar breeding in Canada. In: Maini, J.S. & Cayford, J.H. (eds), Growth and utilization of poplars in Canada. *Canadian Department for Rural Development, Forestry Branch Publication* 1205, pp. 88–100.

Heybrook, H.M. (1978) Primary considerations; multiplication and genetic diversity. *Unasylva* 30 (119–120), 27–33.

Kolster, H.W. (1978) Menging van klonen in populierenbeplantingen. *Populier* 15 (1,2), 27–32.

Libby, W.J. (1980) What is a safe number of clones per plantation. In: Heybrook, H.M., Stephan, B.R. & von Weissenberg, K. (eds), *Proceedings Workshop Genetics Host–Parasite Interactions in Forestry,* Wageningen, September 14–21, 1980, Pudoc, Wageningen, pp. 342–60.

Libby, W.J. (1987) Testing for clonal forestry. *Annales Forestales* 13/1–2, 69–75.

Ontario Ministry of Natural Resources (1983) New forests in Eastern Ontario. *Hybrid Poplar Science and Technology Series* 1, 336.

Weisgerber, H. (1979) The importance of the international *Populus trichocarpa* provenance trial for breeding and cultivation of balsam poplars in the Federal Republic of Germany. In: *Proceedings of the IUFRO Meeting on Poplars in France and Belgium*, September 17–22, 1979, pp. 215–25.

Zsuffa, L. (1975) Some problems of hybrid poplar selections and management in Ontario. *Forestry Chronicle* 51(6), 240–2.

Zsuffa, L. (1976) Vegetative propagation of cottonwood by rooting cuttings. Invited paper. In: *Association of Cottonwood Symposium*, Greenville, Miss, Louisiana State University, Division of Continuing Education, Baton Rouge, Louisiana, pp. 99–108.

Zsuffa, L. (1985) Concepts and experiences in clonal plantations of hardwoods. In: *Proceedings 19th Meeting CTIA, Part 2. Symposium on Clonal Forestry*, Toronto, Canada, Aug. 22–26, 1983, Canadian Forestry Service, Environment Canada, Ottawa, pp. 12–25.

Zufa, L. (1969) Tree breeding and forest genetics in Canada. *Forestry Chronicle* 45(6), 402–8.

Zufa, L. & Zivanov, N. (1966) Indications of significant and specific correlation between the poplar clones and soil types. *Sumarski List* 90(1–2), 137–48.

Chapter 8

Rooting Trials with Stem Cuttings of *Erythrina poeppigiana*: Genotypic and Seasonal Variation in Rooting

Michele C. Rodrick and L. Zsuffa

Faculty of Forestry, University of Toronto,
33 Willcocks Street, Toronto, Ontario, M5S 3B3 Canada

Erythrina poeppigiana (Walpers) O.F. Cook of the family Leguminaceae, sub-family Papilionaceae or Fabaceae, is a nitrogen-fixing tree belonging to a genus containing 110 species of wide distribution in the tropics (Krukoff and Barneby, 1974). The natural range of the species extends from Venezuela and Panama in the north to the western parts of the Bolivian and Peruvian Amazonia (Borchert, 1980). However, it has been introduced and naturalized in various countries in Central America, the Caribbean, and the West Indies where it is used in agroforestry systems (Borchert, 1980; National Academy of Science, 1979).

In Costa Rica (10°N, 84°W), where the present study took place, *E. poeppigiana* is frequently seen growing in tropical zones having an average temperature of between 18°C and 28°C, with annual precipitation of 1500–4000 mm, and from sea level up to 1400 m. It is generally encountered on all types of soils. It grows well on poor and sandy soils, however, preferring deep, clay-type soils.

The tree is grown as a live fence along roadsides, for shade in coffee and cocoa plantations, and associated with pastures in many parts of Central and South America (Budowski, 1984). In these plant associations, *Erythrina* trees regulate light, decrease ambient and soil temperatures, improve water balance and improve soil fertility by way of organic matter shed or pollarded from the trees.

Pollarding is the removal of the crown of the tree to promote branching. In Costa Rica, this is commonly carried out twice a year, just before the coffee harvests, to promote rapid ripening of the coffee berries. The excised branches are spread out among the coffee bushes and left to decompose. In some plantations, they are manually incorporated into the soil after decaying. Decomposition is rapid due to the softness of the wood and the warm, humid microenvironment around the base of the coffee bushes.

Vegetative propagation practices

In Costa Rica, farmers commonly propagate *Erythrina* sp. by excising stakes (1.5 m in length) from established, pollarded trees and inserting them in the soil where they are expected to root and grow. However, *E. poeppigiana* stakes do not root well; they have a variable but high mortality rate (Sanchez, 1986). The farmer frequently has to go back and replace the large number of stakes that did not survive, thereby consuming time and effort.

Previous research has investigated and developed cultural methods to improve the rooting of cuttings of *E. poeppigiana* and other research is presently in progress (Sanchez, 1986). However, the influence of the ortet (parent tree) on rooting behaviour, specifically genetic and seasonally determined physiological differences, has not been examined.

Problem statement

Cuttings have shown a very variable survival (rooting). Up to now, cuttings have been taken at random by farmers and researchers from pollarded trees growing in live fences and from those found within coffee plantations. These cuttings taken from different trees are usually pooled together and subsequent experimentation is carried out on this mixture of clones. Therefore, any differences in individual rooting abilities of ortets, if they existed, would not have been detected. Hence it was important to determine if significant differences exist in the rooting ability of ortets of *E. poeppigiana.*

If significant differences do exist, future breeding work could include ease of rooting as a desirable characteristic. Further research could be carried out on proven difficult or easy-to-root clones and it would be possible to use rooting ability combined with other characteristics as an aid in clonal identification.

Studies have indicated that tropical woody plants undergo periodicity in growth and development, regulated internally but also frequently influenced by seasonal changes in weather patterns. *E. poeppigiana* has been observed to abscind leaves and flower in the dry season (January–April) while displaying several growth flushes during the wet season consisting of new shoot emergence, leaf expansion, and mild leaf abscission

(Borchert, 1980). Considering these two periods in which the tree is in diverse physiological states, it is a distinct possibility that propagules taken from the same tree during these two seasons will behave differently.

In Costa Rica, *E. poeppigiana* cuttings are currently taken throughout the year. However, no observations have been made on rooting behaviour in the two seasons. It would be useful to determine if the physiological states of the ortet in the wet and dry seasons has an effect on the rooting success of *E. poeppigiana* stem cuttings.

Methodology and results of investigation

The study undertaken at CATIE (Centro Agronomico Tropical de Investi-gacion y Ensenanza), Costa Rica, was supported by Canadian International Development Research Centre (IDRC). It was composed of four rooting trials (Rodrick, 1989). The first screened the rooting ability of 15 ortets, each one with different bark characteristics and collected from a different coffee plantation to ensure as much as possible that the genotypes were different. The second used clones from the first trial – two selected for poor rooting ability, one for moderate rooting ability, and one for good rooting ability. Three other genotypes at each site/plantation were selected to test for rooting ability within and between sites. The same ortets were estab-lished in rooting trials in the dry season (January) and again in the wet season (August) to determine the effect of season on rooting success. A fourth rooting trial used ortets from an established provenance trial to test for variation in rooting ability between ortets from widely separated origins.

Significant differences in rooting ability existed between ortets. Clonal range for percent rooted cuttings varied from 0% to 90% for most of the trials. Clonal variation in rooting ability was significant within sites but not between sites. Broad-sense heritability for rooting ability ranged from 0.66 to 0.76 for the trees used in the coffee plantations. The percentage of rooted cuttings was significantly higher for most ortets when cuttings were taken in the dry season compared with the wet season. A marked seasonal difference in rooting success was apparent which coincided with differences in phenology and nutrient status of the ortet.

Conclusions

The following conclusions were reached through these series of experi-ments.

1. Genotype has a significant influence on the rooting ability cuttings.

2. Broad-sense heritability for rooting ability is generally high.

3. Significant differences were apparent between clones in production of root biomass, leaf biomass and root:shoot ratio on newly rooted cuttings.

4. Significant differences exist in the foliar nutrient concentration of ortets between the wet and the dry seasons.

5. A negative relationship exists between foliar nutrient concentration of the ortet and cutting survival.

6. Rooting success was higher in the dry season as compared with the wet season.

7. Sprouting activity was significantly different between the wet and dry seasons.

8. Leaf number and leaf biomass of newly rooted cuttings were significantly higher while root biomass and root:shoot ratio were significantly lower in the dry season.

References

Borchert, R. (1980) Phenology and ecophysiology of tropical trees: *Erythrina poeppigiana* O.F. Cook. *Ecology* 61(5), 1065–74.

Budowski, G. (1984) An attempt to quantify some current agroforestry practices in Costa Rica. In: Huxley, P.A. (ed.), *Plant Research and Agroforestry.* ICRAF, Nairobi, Kenya. pp. 75–92.

Krukoff, B.A. & Barneby, R.C. (1974) Conspectus of species of the genus *Erythrina. Lloydia* 37, 332–459.

National Academy of Sciences, USA (1979) *Tropical Legumes: Resources for the Future.* Washington, D.C., pp. 5–9.

Rodrick, M.C. (1989) 'The influence of genotype and season on rooting of stem cuttings of *Erythrina poeppigiana* (Walpers) O.F. Cook.' Unpublished MSc thesis, University of Toronto, 116 pp.

Sanchez, G.A. (1986) *Project* Erythrina. *Phase I. Final Report.* CATIE/IRDC. 125 pp.

Chapter 9

Rapid Propagation of Fast-growing Tree Species in Developing Countries; its Potentials, Constraints and Future Developments

M. Hosny El-Lakany

Adjunct Professor of Forestry, Alexandria University,
Managing Director, Desert Development Center,
The American University in Cairo

Interest in fast-growing trees, has increased since the early 1960s and gained more momentum after multipurpose species had been declared by the World Bank and FAO as their focus for forestry research for the decade 1985–1995. This has been a natural response to the growing demands worldwide for wood and other forest products.

Several developing countries, especially those with diminishing forest resources rely on multipurpose tree species in their afforestation and reforestation programmes. Tree planting in the tropics and subtropics is increasing despite the disappointing ratio between rates of deforestation and reforestation. It is estimated that nearly 3 million hectares of industrial plantations are established annually in the tropics, often using seeds of inferior quality (Palmberg, 1989). The situation is even worse in arid and semi-arid regions, where meagre tree planting is done with mostly unsuitable species or strains.

Some international efforts to increase the productivity of certain multipurpose tree species, mostly through genetic manipulation, have been outlined by Burley and Stewart (1985). The recommended programmes depend on producing improved seed through conventional tree improvement procedures, with tissue culture as a complementary activity.

This paper deals with the application of biotechnology in the rapid propagation of forest trees, with special emphasis on constraints and potentials in developing countries.

Biotechnology in forest tree propagation

Commercial tree propagation depends on traditional methods, i.e. seeds and to a lesser extent on rooting of stem cuttings. Recently however, certain biotechnologies have been implemented for producing planting stock, though on a small scale. The most common biotechnologies in use include tissue and organ culture, somatic embryogenesis and micrografting. *In vitro* cultures have occasionally been used for early screening for tolerance/resistance to pathogens and environmental stresses in some forest trees as well.

While such traditional biotechnologies have been applied only to a limited extent in forestry, more advanced biotechnologies such as protoplast fusion, haploid cultures, gene cloning, gene transfer, and recombinant DNA are in the early stages of development in general and still on a very limited research scale in forestry, (Bonga *et al.*, 1988, and Palmberg, 1989 unpublished).

Tissue culture as a tool for advancing clonal forestry has been used quite successfully for a number of fast-growing species. Notable success has been achieved with some poplar hybrids (Whitehead and Giles, 1976) and in *Paulownia* (Radojevic, 1979). Brown and Sommer (1982) list forest tree species that have been propagated through various tissue culture techniques, including some fast-growing, multipurpose tree species belonging to the genera: *Populus, Eucalyptus, Acacia, Bombax, Dalbergia, Tectona, Platanus, Casuarina, Gmelina, Ficus, Shorea, Terminalia* and *Salix*.

The work in Australia on screening for and developing of salt-tolerant lines of *Eucalyptus camaldulensis* (Hartney, 1982) is noteworthy. Some lines are now available commercially. Clonal propagation of elite trees of *Casuarina glauca, C. cunninghamiana* and *C. equisetifolia* has been recently achieved in Egypt (El-Lakany and Barakat, in preparation).

Because of recent success in mass propagating young and, to some extent, mature trees by shoot tip or axillary bud cultures and the potentialities this approach offers to commercial production of plantable tree seedlings, many research and development efforts have been devoted to this purpose. Several excellent reviews are available giving detailed accounts of research leading to organogenesis of woody trees and the potentialities such techniques offer in tree improvement and production forestry programmes (cf Brown and Sommer, 1982).

Progress has not been easy because of many problems peculiar to forest trees, such as difficulties in rejuvenating mature tissues, exudates that inhibit growth *in vitro*, ... etc. There are some indications that cells and tissues associated with sexual reproduction in older trees such as nucellar tissue, ovules, or somatic anther cells, may be induced to undergo organogenesis more easily than other parts of the same plant. For example, Duhoux *et al.* (1986) used immature female inflorescences to micropropagate *Casuarina equisetifolia* in order to produce homogeneous material needed for the study of actinorhizal symbiosis.

Basal root suckers or epicormic shoots provide a good source of material for initiating cultures. Although plantlets have been produced frequently from callus cultures, the practical use of this technique is restricted to species or cultivars capable of high incidence bud differentiation and subsequent shoot growth (Brown and Sommer, 1982). There are several reports in the literature of successful embryogenesis of some woody dicotyledons in tissue culture, but only a few have been developed to commercial utility in forestry.

While the best silvicultural and managerial methods must be practised in the afforestation and reforestation programmes implemented in developing countries, the use of appropriate germplasm is even more important. The research needs in forest tree breeding and improvement in developing countries were outlined by Palmberg (1989). It appears that realization of practical results is not an easy task.

Even in North America and Europe where very active and progressive tree improvement and forest genetics programmes are supported by governments, industry and research institutions, insignificant amounts of genetically improved seed are available for reforestation. In Canada, for example, for the ten major reforestation species only 0.02% of the nearly 15 billion seeds collected for regeneration purposes come from seed orchards that will yield genetically improved stock. Such a low percentage is expected to reach only 1.6% by 1997 (Cheliak, 1987). One basic problem that faces all conventional tree improvement programmes is the time lag between programme initiation and delivery of genetically improved propagules to the field. A model programme in which micropropagation and/or somatic embryogenesis are integrated into conventional tree improvement schemes in order to overcome some of the problems associated with time lag has been described by Cheliak (1987). It should be applied in developing countries.

Constraints

Rapid propagation of fast-growing trees in developing countries faces certain technical, human and financial constraints. Technical problems of biotechnology in forest trees which are related to the nature of the material used are common to industrialized and developing countries. Thus, research and development are continually dealing with such issues and advancement in science should eventually be able to overcome most of these problems. It should be borne in mind that a large amount of research is being done by private enterprises and although theoretically there should be no barriers for the transfer of the knowledge obtained developing countries should be prepared to pay for acquiring certain technologies.

Some developing countries suffer from a shortage of infrastructure, equipment, access to information, specialized personnel and interdisciplinary teams of scientists conversant with these new technologies (Palmberg, 1989, unpublished). There may also be a lack of integration between conventional tree breeding and biotechnology research programmes.

Another technical problem that faces developing countries is that most of the progress in forest tree biotechnologies has been achieved with species that are not on their priority lists, perhaps with the exception of some *Eucalyptus* and *Acacia* species. Although the basic procedures may be comparable among many species, the important fast-growing multi-purpose tree species may require specific protocols. Substantial fundamental research, which many developing countries cannot afford, would therefore be needed. Since many results are documented in the international and local literature, information transfer between industrialized and developing countries as well as among developing countries themselves needs to be strengthened and promoted.

Training of potential researchers in industrialized countries is of utmost importance for the development of biotechnologies in developing countries. In this rapidly developing field, refresher training of senior researchers is essential, as recommended by the FAO/CTA symposium on plant biotechnologies for developing countries (Palmberg, 1989, unpublished). It is highly desirable to strengthen certain existing research institutions in developing countries rather than to initiate new ones.

Financial constraints are major aspects hindering the development of biotechnologies in developing countries. While some of these technologies may be economic in industrialized countries, it is not expected that the situation would be similar in developing countries unless activities are confined to clonal propagating of highly valuable genotypes using the simplest technology possible.

The largest cost in micropropagation is the labour and time entailed with manual subculturing techniques (Hartney and Kabay, 1984). Such a constraint may be turned into an advantage in developing countries if readily available technicians are employed to do this work after simple training.

Future prospects

With unprecedented progress in biotechnology during the past two decades, coupled with the continually increasing demand for wood and tree products throughout the world, one can predict with certainty that asexually produced trees will comprise a significant portion of forest planting stock in the future. While propagation by rooted cuttings will continue in the future for some species such as hybrids of poplars, eucalypts and others, micropropagation will gain more ground because of three main advantages: much higher multiplication rates; a greater degree of control; and the small space required. There are several methods available to reduce the cost of producing plants by micropropagation and there is potential for integrating tissue culture techniques into nursery systems developed for the production of seedlings (Hartney and Kabay, 1984).

The production of embryoids in tissue culture has several advantages over organogenesis as a practical means of propagation. It is expected that this technology may make mass cloning of tree species operational and economically feasible (Brown and Sommer, 1982). Fast-growing, multi-purpose tree species deserve a special consideration in this respect.

Among the many proposed applications of biotechnology to plant breeding in general, those concerned with screening and segregating material are perhaps the most important for wide application in the near future (Arnold, 1987). A potential area that would be immensely important in forestry is the use of somatic embryogenesis in screening for tolerance to environmental stresses. For example, large areas of salt-affected soils have been recommended to be used for fuelwood and fodder production (El-Lakany, 1986). Thus, rapid cloning of salt-tolerant, fast-growing strains will eventually be important. Similarly, breeding for drought and disease resistance could be feasible using this technique.

The recent development of techniques to obtain haploid cells or proto-plasts through androgenesis or gynogenesis *in vitro* has opened new avenues to detect and select cells or protoplasts with useful genes, e.g. for resistance to a specific phytotoxin, for salt tolerance or for drought resist-ance, if these are expressed at the cellular level. Among other possibilities listed by Bonga *et al.* (1988), is the production of interspecific somatic hybrids to bypass sexual barriers.

Research and development in multiple use agroforestry systems coupled with simple biotechnology, where scientific and labour manpower is readily available, may offer the best opportunity and short-term profit-ability in developing countries (Durzan, 1982).

In addition, the vital role of biotechnologies in the exchange of germ-plasm and gene conservation should not be overlooked in developing countries.

Conclusions

Rapid propagation of fast-growing, multipurpose trees on a large scale has been achieved through simple tissue and cell culture in only a few species. Such technologies are expected to be both technically and economically feasible in many developing countries despite some constraints. Certain species of *Acacia*, *Casuarina*, *Eucalyptus*, *Paulownia* and *Populus*, offer the best examples.

Tree breeding programmes which incorporate biotechnology for cloning superior genotypes and for early screening for stress tolerance may be the most economically and technically feasible options for developing countries. Micropropagation or somatic embryogenesis may be employed to reproduce specially selected (elite) trees and/or the progenies of the best families (half and full sibs) in seed orchards.

Even in the US, it became apparent very early that advanced biotechnology would be an expensive, high risk technology, and one that has to be entered into with care and a high degree of public sensitivity (Krugman, 1986). Therefore, developing countries should, for the time being, refrain from directing their limited resources towards research and development of new biotechnologies such as genetic engineering and parasexual hybridization of forest trees.

References

Arnold, M.H. (1987) The application of molecular biology to plant breeding. In: Rao, A.N. & Mohan Ram, Hy. Y. (eds), *Agricultural Application of Biotechnology, Proceedings Nayudamma Memorial Symposium.* COSTED, India, pp. 4–16.

Bonga, J.M., von Aderkas, P. & James, D. (1988) Potential application of haploid culture of tree species. In: Hanovera, J.W. & Heathley, D.F. (eds), *Genetic Manipulation of Woody Plants.* Plenum, New York, pp. 57–77.

Brown, C.L. & Sommer, H.E. (1982) Vegetative propagation of dicotyledonous trees. In: Bonga, J.M. & Durzan, D.J. (eds), *Tissue Culture in Forestry.* Martinus Nijhoff/Dr W. Junk, The Hague, pp. 109–49.

Burley, J. & Stewart, J.L. (1985) *Increasing Productivity of Multipurpose Species.* IUFRO, Vienna, 560 pp.

Cheliak, W.M. (1987) Biotechnology in tree improvement. In: Galloway, R.L., Greet, R.M., McGowan, D.W.J. & Walker, J.O. (eds), *Primeval Improvement: The New Forestry, Agriculture Proceedings of Symposium* COJFRC 0-P-15. pp. 57–61.

Duhoux, E., Sougoufora, B. & Dommergues, Y. (1986) Propagation of *Casuarina equisetifolia* through axillary buds of immature female inflorescences cultured *in vitro. Plant Cell Reports* 3, 161–4.

Durzan, D.J. (1982) Cell and tissue culture in forest industry. In: Bonga, J.M. & Durzan, D.J. (eds), *Tissue Culture in Forestry.* Martinus Nijhoff/Dr W. Junk, The Hague, pp. 36–71.

El-Lakany, M.H. (1986) Fuel and wood production on salt affected soils. In: Barrett-Lennard, E.G., Malcolm, C.V., Stern, W.R. & Wilkins, S.M. (eds), *Forage and Fuel Production from Salt-affected Wasteland.* Elsevier, Amsterdam, pp. 305–18.

Hartney, V.J. (1982) Tissue culture of *Eucalyptus. Proceedings International Plant Propagators Society* 32, 98–109.

Hartney, V.J. and Kabay, E.D. (1984) From tissue culture to forest trees. *Proceed-*

ings International Plant Propagators Society 34, 93–9.

Krugman, S.L. (1986) U.S. Forest Service's view of recent developments and the future of forest biotechnology. In: *TAPPI Research and Development Conference Proceedings*, pp. 79–91.

Palmberg, C. (1989) Research needs in forest tree breeding and improvement in developing countries. *Agroforestry Systems* 9(1), 29–35.

Radojevic, L. (1979) Somatic embryos and plantlets from callus cultures of *Paulownia tomentosa* steud. *Zeitschrift für Pflanzenphysiologie* 91, 57–62.

Whitehead, H.C.M. and Giles, K.L. (1976) Rapid propagation of poplars by tissue culture methods. *Proceedings International Plant Propagators Society* 26, 340–3.

Chapter 10

Development of High-yielding Clonal Plantations of *Eucalyptus* Hybrids in the Congo

Oudara Souvannavong

Programme Amélioration du Matériel Végétal, Centre Technique Forestier Tropical, Départment du CIRAD, 45bis, avenue de la Belle Gabrielle, 94736 Nogent-sur-Marne Cédex, France

The Congo is an equatorial country of Africa, with a total area of 342,000 km^2 and 1.9 million inhabitants. Sixty percent of the country is covered with natural forests which provide a significant part of the national income. Timber export volume was 715,000 m^3 in 1986.

Since the end of the Second World War efforts have been made to balance the country's wood production with man-made forests of timber, and fast-growing species. Several experimental projects have been initiated. The main one concerns *Eucalyptus* plantations on the coastal savannah surrounding Pointe Noire which is the only port and the second city of the Congo. The objective was originally the fuelwood supply of Pointe Noire. The main scope is now to produce pulpwood and poles, fuelwood being a by-product. The very fast development of large scale high-yielding clonal plantations is a good illustration of the important impact that rapid mass vegetative propagation techniques can have. The main steps are given in Table 10.1.

Natural conditions

Pointe Noire is located on the West coast of the African continent at a latitude of 4°45'S. The mean annual temperature is 24.5°C with little seasonal variation. The mean annual rainfall is 1250 mm with one distinct

Table 10.1. Main steps of clonal forestry development in Congo, based on rapid propagation of *Eucalyptus* hybrids.

1956 onwards: Species selection
Introduction of 63 *Eucalyptus* species

1963: Observation of outstanding hybrids in progenies collected from trial plots (natural pollination)

1970: Establishment of Bi-specific seed orchards with species producing good hybrids

1974: Success in rooting cuttings (rooting *Eucalyptus* was reputed impossible until then)

1975: First clonal tests

1978: Start of large scale plantations with high-yielding hybrid clones

1978 onwards: Use of controlled pollination to create new interspecific combinations

1987: More than 23 000 ha of plantations
Planting of 25 000 more hectares
Start of harvest

---

dry season, from November to May, characterized by cool temperature, cloudiness and high air humidity.

The savannahs cover low sandy plateaux (30–70 m in altitude). The natural vegetation is composed of grass and sparse scrub. The soil is deep but very poor. These areas are not used for agriculture or cattle breeding. Experimental plantations have been started to assess the possibility of utilization for forest plantations.

Species and provenance selection

A large array of fast growing species has been introduced in preliminary tests. *Eucalyptus* species rapidly turned out to be the most suitable. Emphasis was consequently put on these species.

Since 1956, 63 *Eucalyptus* species (350 provenances) have been put into trials. Eighteen species have been selected for their adaptation, their productivity and subsequently their interest as hybrid parent species:

1. Species which are well adapted for many provenances
- productive   –   *E. urophylla, E. cloeziana*
- fairly productive –   *E. tereticornis, E. brassiana*
- poorly productive   –   *E. alba, E. torelliana*

2. Species which are well adapted for a few provenances
- productive   –   *E. pellita, E . resinifera, E. pilularis*
- fairly productive –   *E. brasii*
- poorly productive –   *E. exserta, E. nesophila, E. raveretiana*

3. Poorly adapted species which are interesting for the productivity of their interspecific hybrids
- productive   –   *E. grandis, E. saligna, E. robusta*
- fairly productive –   *E. citriodora*

Clonal utilization of interspecific *Eucalyptus* hybrids

In 1963, outstanding individuals were observed among trees raised from seeds collected in species trial plots. These were assumed to be interspecific hybrids resulting from cross pollination, by natural agents, of neighbouring species in the trials. These 'natural' hybrids are of two kinds. The first one, originally named *Eucalyptus* PF1, is a cross between *E. alba* and a group of hybrids with *E. grandis, E. robusta, E. urophylla* and possibly *E. botryoides* ascendants. The other one, called *E.* 12ABL × *E. saligna* is actually a *E. tereticornis* × *E. grandis* cross.

Bi-specific seed orchards were established in 1970 to produce hybrid seeds. The results of this operation were disappointing because seed production was uneven and the hybrid progenies were very heterogeneous. One could not rely on such a supply of hybrid plant material for large scale plantations. Research was therefore started in 1972 that led, in 1974, to the development of a reliable rooting technique that allowed the vegetative propagation of selected hybrid trees. The rooting technique has been progressively improved. The first clonal tests were planted in 1975 and the first large scale pilot plantation project (3000 ha) started in 1978. A major improvement in the process was obtained in 1982 when production time in the nursery was brought down from 6 months to 8 weeks. The annual rate of planting was then increased (up to 5000 ha some years).

The rooting operation sequence is as follows:

1. coppicing of the multiplication garden (15–30 cm above ground) 1 month before the start of the rainy season;
2. collection of the shoots 6–8 weeks after;
3. preparation of two-leaf cuttings, fungicide and hormone applications, base rolled in a flat tissue bag filled with vermiculite and fertilizers;
4. 3 weeks under daytime mist (with 60% shade during the first 3 days) for rooting;

5. 3 weeks of weaning by progressive reduction of the time of mist application;
6. selection;
7. 2 weeks of hardening outside mist;
8. planting.

Because of the very early start of large scale plantations, not all the clones used during the first 2 years were tested clones. But since 1980 only clones having gone through a clonal selection scheme are used. This selection scheme comprises:

1. individual selection in hybrid progeny plots;
2. vegetative propagation by cuttings;
3. first stage clonal tests (unreplicated monoclonal 50 ramet plots);
4. selection (growth, stem form);
5. second stage clonal tests (replicated monoclonal plots);
6. selection (growth, stem form, pulp yield);
7. mass propagation for large scale plantation.

Each year new clones are put into use. To minimize the sanitary hazards the plantation layout is a patchwork of monoclonal compartments of 25 ha each. Each year 10–20 different clones are used. The clones usually differ for two following years. The total number of clones in use is around 60. The increase in yield obtained is important.

In 1970, the yield of the best provenances of the best species (*E. tereticornis* and *E. urophylla*) was 12 m^3 ha^{-1} y^{-1}.

In 1987, the yield of the clonal plantations ranged from 20 to 25 m^3 ha^{-1} y^{-1}, the best clones yielding 30–38 m^3 ha^{-1} y^{-1}. The pulpwood is exported to Scandinavia and Southern Europe.

Breeding programme

The genetic base of the 'natural' hybrid clones is very narrow:

1. the PF1 clones are almost all from only two distinct mother trees;
2. the '12ABL' origin of the other hybrid type is also very narrow (possibly one tree).

The creation of new interspecific hybrids began in 1978 using controlled pollination techniques. The objectives are to create more productive clones and to enhance the genetic diversity of the clones used in the large scale plantations.

From 1978 to 1983 an extensive testing of possible interspecific combinations was done to find out which are the more interesting. Forty-four different combinations were studied among which were selected:

E. alba × *E. grandis*;
E. urophylla × *E. grandis*;
E. tereticornis × *E. pellita*;
E. urophylla × *E. pellita*;
E. urophylla × *E. resinifera*.

Since 1983 cross-breeding plans are being undertaken for the more interesting combinations.

A total of 285 artificial hybrid clones have been included in the clonal tests from 1982 to 1987.

Conclusion

The development by the Centre Technique Forestier Tropical/Congo of the rooting technique for *Eucalyptus* trees has led to a fast and important use of high-yielding hybrid clones in large scale plantations by the Unité d'Afforestation Industrielle du Congo (UAIC). This success is a good illustration of the power of mass propagation techniques but this should not overshadow the importance of proper selection and breeding programmes which are necessary to produce, in the medium and long terms, the genotypes to be mass propagated.

Bibliography

Chaperon, H. (1978) Breeding improvement of hybrid *Eucalyptus* species in the Congo Brazzaville. In: Nikles D.G., Burley J. & Barnes R.D. (eds), *Progress and Problems of Genetic Improvement of Tropical Forest Trees*. Oxford Forestry Institute, Oxford, pp. 1027–39.

Delwaulle, J.C. (1989) Plantations clonales au Congo. Point des recherches sur le choix des clones dix ans aprés les premières plantations. In: Gibson G.L., Griffin A.R. & Matheson A.C. (eds), *Breeding Tropical Trees: Population Structure and Genetic Improvement Strategies in Clonal and Seedling Forestry*. Oxford Forestry Institute, Oxford, UK and Winrock International, Arlington, Virginia, USA, pp. 431–4.

Vigneron, Ph. (1989) Les hybrides artificiels d'*Eucalyptus* au Congo, création et multiplication. In: Gibson G.L., Griffin A.R. & Matheson A.C. (eds), *Breeding Tropical Trees: Population Structure and Genetic Improvement Strategies in Clonal and Seedling Forestry*. Oxford Forestry Institute, Oxford, UK and Winrock International, Arlington, Virginia, USA, pp. 425–7.

Chapter 11

Positive Implications of New Biotechnology for Asian Forestry

Darus Haji Ahmad

Forest Research Institute, Kepong,
52109 Kuala Lumpur, Selangor, Darul Ehsan, Malaysia

In Asia intensive and improper logging activities and conversion of forest lands for agriculture and other non-forest use have led to large areas of tropical forest being destroyed. At today's exploitation rate of 7.3 to 11 million hectares annually and with less than 10% reforestation, there will not be much forest area left and all productive forests may be lost before the turn of this century (Rao and Lee, 1982; Burley, 1987). This will have important effects on the environment and on overall economic development. For example, Malaysia's export earnings for timber and timber products was about $6880 million or about 15.2% of the country's export earnings in 1987 (Anon., 1988). Successful restoration of the forests after logging is indeed limited and the recovery and re-establishment effort at forest regeneration leaves much to be desired. The gap, termed by Brodie and Tedder (1982) as 'regeneration delay', appears to be widening. With a slow reforestation rate, the implication on future resource flows is obvious.

Hidden in the already delayed regeneration are silvicultural problems. In the case of Malaysia and in some other countries these include:

1. low regeneration success rate as exemplified for example by Tang (1980), and Salleh and Johari (1983);
2. high costs, with doubtful economic viability of regenerating the forest as indicated by Rauf (1983, unpublished);
3. slow growth of species regenerated.

Intensive silviculture is still a primary concern in forest management, although Gillis *et al.* (1983) concluded that until very recently, tropical timber was typically considered to be a renewable resource. However, several factors have led many forest specialists to classify this resource as non-renewable, or at best semi-renewable.

If historical patterns are any indication, an important part of forest products in the future will come from plantations as in much of agriculture. This is in fact already occurring in many Asian countries. Malaysia for instance, embarked on an ambitious and intensive forest resource development programme by establishing forest plantations under the compensatory Plantation Programme. Under this programme, 188,200 ha of fast-growing, high-yielding species, such as *Paraserianthes falcataria, Gmelina arborea, Acacia mangium* are to be grown for general utility lumber. Partly funded by the Asian Development Bank (ADB), the estimated cost of establishment is $24.5 million over a 7 year period. With such a conversion to forest plantations, future timber stands will be similar to those normally found in the northern temperate countries.

One of the major decisions when creating forest plantations is the type of tree species to be planted: whether to use indigenous or exotic species. The introduction of endogenous species in forest plantations has added to the nature–nurture imbroglio, although the experience in the use of indigenous trees as plantation species is very limited, except for several scattered experimental plots.

The rapid exploitation rate brings with it the problem of the depletion of gene resources of both trees and shrubs in natural forests. There is an urgent need to improve conservation techniques for woody species. At present, the conventional conservation strategies of tree species, such as natural forest reserves, forest plantations, seed stands, arboreta, and clonal banks are being further developed and maintained. Such attempts are being made in order to prevent the loss of wild gene resources of tree species, to maintain environmental stability, to provide sanctuary for wildlife as well as to ensure a sustained supply of timber for the future. Such storehouses of genetic resources will be useful for the improvement of tropical tree species, agricultural crops and livestock.

However, forest plantation programmes face many problems, in particular the inadequate supply as well as poor quality of seedlings. Thus, it seems possible that biotechnology, which has already been successful in agriculture, will play a very important role in forestry, especially in conservation and mass production of genetically improved, superior planting materials for reforestation.

Biotechnology and forest development

There are four biotechnological areas currently under rapid development in agriculture: 1) genetic engineering; 2) cell and tissue cultures; 3) enzyme technology; and 4) bioprocess engineering. However, not all of these techniques are of immediate relevance to forestry. In this paper, the main focus will be on the possible application of cell and tissue culture technologies for the forestry sectors in Asian countries.

Tissue culture research on woody plants has been limited because of the difficulties of propagation, especially at the stage of isolating uncontaminated cultures and inducing root and shoot differentiation from either callus or organ explants. However, recent successes with cultured mature teak, acacias, eucalypts, poplar and pines, have renewed the interest in the use of tissue culture and in its potential for clonal forestry (Mascarenhas, 1984).

The inadequate supply as well as poor quality of seedlings have been reported in most reforestation projects in Asian countries. The need for improved and genetically superior planting materials and the production of large quantities of stock plants for reforestation will be substantial. We believe that biotechnology can make important contributions in the areas of: 1) producing genetically superior planting materials; 2) mass production of selected planting materials; 3) germplasm conservation; and 4) improving the suitability of planting materials for various locations.

Tree improvement and breeding

In vitro propagation methods by protoplast cultures and genetic engineering of economic agriculture crops for the purpose of mass production and hybridization hold great promise in the breeding and production of forest tree species. However, unlike annual crops, tree species have proved slow to improve genetically, because of their long gestation period. These new methods can, however, be easily applied to tree species as a first step. For example, Darus (1988) has been successful in the direct mass production of genetically superior *Acacia mangium* clones through normal micropropagation techniques.

Another method of tree improvement and breeding is genetic manipulation. Although this is a relatively new development in woody species it has been extensively applied in the agricultural sector and has already created a new generation of important crop plants which produce greater yields, have better nutritional characteristics and herbicide resistance, or have a broader range of tolerance to drought and other environmental stresses. The question now arises, whether the same techniques can provide similar benefits to forestry. For forestry, manipulation at the DNA level may be the best way to improve performance as well as genetic variability in a short period of time. If after successful manipulation at the DNA level, the tissue can be cultured, the treelets produced could be clonally propagated and their offspring could be field tested in about one-third of the time required in normal breeding cycles.

However, there are barriers that may require time to overcome (Trotter, 1986). A major problem encountered in the application of DNA techniques to trees is the lack of basic information, especially on the biochemistry and physiology of tree species, and this has hindered progress towards the control and quality of tree growth (Trotter, 1986). Governments, research institutions as well as the private sector should co-operate in the application of these new biotechnologies to plant breeding. In Japan, for example, private industries and the government have recognized the importance of plant biotechnology as a future fertile ground for the modern seed industry and clonal forestry.

Mass production of stock plants

The difficulties in reforestation of tropical species are related to the nature of seed production as well as the viability of the seeds. Unpredictable flowering and seeding, and harvesting of large quantities of seeds are becoming of increasing importance in tropical forestry (Sasaki and Rahman, 1978). Moreover, due to the recalcitrant nature of the seeds, serious storage problems arise. Production of stock plants of tree species, especially tropical ones, by stem cuttings has so far proved difficult (Burley, 1987). There is now an urgent need to improve techniques for the production of genetically improved stocks of important forest species for planting programmes. Plant cell and tissue cultures may provide alternative methods to fulfil this requirement and may also contribute to germplasm conservation.

Cloned plants which are genetically identical to the original donor plants could be produced by cell and tissue culture techniques. By propagating selected genetic material, the immediate gains can be expected to be greater than by the production of seeds in traditional seed orchards. On uniform sites, the individuals of each clone would be very similar in appearance to the original mother trees. Production should be developed of the best clones direct from selected trees with superior timber quality, volume production and resistance to factors such as fungal or bacterial diseases and other pests. This would mean decreasing plantations of wild populations that are genetically undefined.

In vitro forest germplasm and conservation

The rich diversity of species in tropical rainforest is well known. Poore (1964) notes for example that tropical moist forests in southeast Asia contain more than 100 tree species per hectare. This rich diversity implies that no single species dominates the community, and that individual species occur in low densities (Furtado, 1979). In peninsular Malaysia alone, the flora is estimated to comprise 10 000 species of which 4100 are woody. About 2900 species reach a diameter of 10 cm d.b.h. (diameter at breast height) while about 1680 species in 375 genera are trees. A 10 ha survey at Pasoh Forest Reserve (a lowland dipterocarp forest) identified a total of

5907 trees of 10 cm d.b.h. and larger belonging to 460 different species. The Dipterocarpaceae constitute about 7–14% of the total number of trees while Euphorbiaceae account for about 6–16%.

The gene pool of forest species as well as many cultivated plants, especially the edible fruit trees and endemic species that exist within the forest should be conserved. Presently, the *ex situ* conservation programme is implemented through the establishment of arboreta, seed banks or plantations: these have their limitations because seed production of tropical forest trees is irregular and erratic and in many instances occurs rarely. Although the rooted cuttings can be used, some species have very peculiar characteristics, such as being plagiotropic: there are difficulties in propagating from mature trees and some rooted cuttings of the older trees may be less adaptable to the soils and other physical conditions in their new location, and offer less resistance to harmful pathogens (Rao and Lee, 1982). For these reasons, the culture techniques appear to be more suitable for mass production and gene conservation. Clonal banks of tree germplasm will be useful for gene conservation of many tropical species. The germplasm may be conserved by two culture-based techniques.

1. *Repeated subculturing of cultured explants.* This technique is considered safe and economic compared with the traditional methods of maintaining clonal materials of selected plants in the nursery or field plots which are not only extremely expensive but there is also the risk of the material being lost as a result of environmental hazards. However, in cultures the relatively small space required as well as the simple techniques can be advantageous for the preservation of a large number of clonally multiplied plants. Additionally, the plants can be maintained in an environment free from pests, pathogens and viruses. For example, in agriculture, strawberry plants stored at 4°C without a subculture were fully viable after 6 years.

2. *Cryopreservation.* In this technique the cultured tissues or organs are stored at very low temperature, (-196°C) in liquid nitrogen (Mascarenhas, 1984). Theoretically, all plants can be stored indefinitely if their metabolism can be completely stopped and restarted when required. To date eight woody species have been successfully preserved (Burley, 1987). This concept has been successfully proven in the preservation of the embryos of tropical recalcitrant seeds. Recalcitrant seeds are normally sensitive to extremes in temperature, need storage at temperatures below 4°C with controlled moisture content, and under normal conditions have a very short life span.

Issues related to positive implications of biotechnologies to forestry

Cost and returns

The use of micropropagation techniques for mass production of selected material with specific genetic traits such as superior wood quality, form, growth rate, pest and disease resistance can be a cost-effective means of establishing viable forest plantations and regenerating forest. Such techniques overcome the main problems using traditional reforestation strategies and particularly the inability to meet the requirement for seedlings to cope with the pace of logging. Although utilization of *in vitro* plant propagation can be labour-intensive compared with conventional methods, reduced labour input can be achieved by eliminating some of the unnecessary steps involved in the production process (Hartney and Svensson, Chapter 2 this volume). For example, the rooting process can be done under *in vitro* conditions and hence the higher cost of production can be eliminated.

The prevailing considerations of the costs of benefits of such projects need to be modified by resource managers. The reason is that conventional assessments evaluate the potential net benefits based on the difference between reforestation gains without biotechnology and with biotechnology. Such simple calculations must take into account all the positive and negative aspects of forest management viewed from a national perspective.

Employment creation

If biotechnology is used effectively to help ensure the viable establishment of forest plantations the forest production does not itself create much employment per unit of natural forest area. Much of this low level of employment opportunities created at the primary level is attributable to the low volume of output from natural forests. However, as the output from the forests is increased, a substantial degree of employment linkages are created at the secondary and tertiary levels and as the forests are ultimately converted into forest plantations, the primary output level will also grow. This would directly impact upon the employment generating capacity of the forests. It is therefore important that viable forest plantations utilizing advances in biotechnologies are established to allow for a much greater contribution of forestry to employment.

Environmental benefits

As mentioned earlier, many tropical Asian forests may be in transition to even-aged forest. This should be the natural choice despite the caution that conversion of complex tropical moist forest into less complex silvicultural and agricultural ecosystems may result in higher albedo (fraction of incident radiation reflected by a surface) and runoff, with consequent loss of nutrients and soil from the system, and in lower rates of forced convection. At the same time, the variety of the living biomass of the area is drastically reduced and with this its capacity to store nutrients and energy and to provide living space for plants and animals.

Beddington and May (1977) state that harvesting of natural resources has tended to ignore the environmental aspects and to treat them as constants. Any disregard for these and other non-pecuniary aspects of forestry must be seen as narrow-minded and short-sighted. The forest resources must therefore be managed not solely for timber, but also for all other benefits. Environmental problems may be reversible or irreversible, more or less certain, temporary or cumulative, independent or synergetic and short and long in the time-scale between cause and effect. The order of gravity may extend from losses of environmental amenities to problems that seriously threaten human health, genetic stock and the sustaining capacity of entire ecological systems. It is essential that such factors be taken into consideration in any forestry management programme.

Conclusion and recommendations

Biotechnology could halt the rapid increase in logging of natural forests by successful implementation of viable plantation forestry and simultaneously conserve the environment. The successful application of biotechnology, including its commercialization, would depend on several factors. First, there must be an adequate research base with a free flow of ideas and technical information. This requirement covers scientific manpower, research facilities and funding and a technical information system. The importance of creating a strong scientific and biotechnological base in terms of infrastructure and manpower to transform the latent potential of clonal forestry into realistic productivity output in developing countries cannot be ignored. In developed countries with a strong research tradition, the universities, the government and private research agencies fulfil this role. For Asian countries, a network of co-operation between scientists and the establishment of appropriate biotechnological centres is required. In addition, the technical capability must be developed so as to transform potentially successful ideas beyond laboratory research into production.

References

Anon. (1988) *Profile of the Primary Commodity Sector in Malaysia.* Ministry of Primary Industries, Malaysia.

Beddington, J.R. & May, R.M. (1977) Harvesting a natural population in a randomly fluctuating environment. *Science* 197, 463–5.

Brodie, J.D. & Tedder, P.L. (1982) Regeneration delay: economic cost and harvest loss. *Journal of Forestry* 801, 26–8.

Burley, J. (1987) Application of biotechnology in forestry and rural development. *Commonwealth Forestry Review* 66(4), 357–67.

Darus, A. (1988) 'Vegetative propagation of *Acacia mangium* by stem cuttings and tissue culture techniques.' Unpublished PhD thesis, University of Aberdeen.

Furtado, J.I. (1979) The status and future of tropical moist forest in Southeast Asia. In: McAndrews, C. & Sien, C.L. (eds), *Developing Economics and the Environment: The Southeast Asian Experience.* McGraw Hill, Southeast Asian Series, Singapore, pp. 73–120.

Gillis, M., Perkins, D.H., Roemer, M. & Snodgrass, D.S. (1983) *Economics of Development.* W.W. Norton and Company, New York.

Mascarenhas, A.F. (1984) Test-tube forests. *Science Today* 20–3.

Poore, M.E.D. (1964) Vegetation and flora. In: Gungwu, W. (ed.), *Malaysia: A Survey.* Pall Mall Press, London, pp. 44–54.

Rao, A.N. & Lee, S.K. (1982) Importance of tissue culture in tree propagation. In: Fujiwara, A. (ed.), *Proceedings of the 5th International Congress on Plant Tissue and Culture.* Tokyo, Japan, pp. 715–18.

Salleh, M.N. & Johari, B. (1983) Silvicultural practices in Peninsular Malaysia. Paper delivered at the *IUCN Symposium on Future of Tropical Rainforest in Southeast Asia.* Kepong, Malaysia.

Sasaki, S., Tan, C.H. & Adb. Rahman, Z. (1978) *Physiological Study on Malaysian Tropical Rain Forest Species.* The Forest Depart. Pen. Malaysia, Kuala Lumpur.

Tang, H.T. (1980) Factors affecting regeneration methods for tropical high forest in South East Asia. *Malaysia Forester* 43(3), 469–80.

Trotter, P.C. (1986) Biotechnology and economic productivity of commercial forests. *Tappi Journal,* 22–8.

Chapter 12

Recommendations

The participants agreed unanimously that all methods of rapid and mass propagation are an important part of the general tree improvement programme and that the collection, conservation and utilization of genetic resources of woody species and access thereto are also essential components of international tree improvement programmes.

1. It was agreed that more research into the development of reliable, inexpensive methods for mass and rapid propagation are required and that *in vitro* propagation methods should only be used if more effective than other methods.

2. It was recommended that there should be more co-ordinated studies of the conservation, and even collection, of seed in some species.

3. It was recommended that in addition to the basic problem of the development of training and educational programmes, especially for Third World countries, and of improved facilities, long-term support for tree improvement programmes needs to be secured.

4. With respect to *in vitro* propagation it was recommended that more basic research into the biochemistry of organogenesis and into the physiology of root and shoot initiation should be developed with increased co-operation between researchers concerned primarily either with fundamental or applied research. In particular the need was stressed for an increased understanding of the roles of nutrients, growth regulators and other constituents of culture media.

5. It was suggested that the proceedings of the meeting should include *inter alia* a review of the advantages, disadvantages, constraints and benefits of the various macro- and micropropagation techniques with specific examples for different species and biomes.

Index